Richard Blum

Sams Teach Yourself Arduino™ Programming in 24 Hours

SAMS 800 East 96th Street, Indianapolis, Indiana, 46240 USA

Sams Teach Yourself Arduino™ Programming in 24 Hours

ISBN-13: 978-0-672-33712-3
ISBN-10: 0-672-337126

Library of Congress Control Number: 2013955616

Printed in the United States of America

First Printing: September 2014

Trademarks

All terms mentioned in this book that are known to be trademarks or service marks have been appropriately capitalized. Sams Publishing cannot attest to the accuracy of this information. Use of a term in this book should not be regarded as affecting the validity of any trademark or service mark.

Arduino is a registered trademark of Arduino and its partners.

Warning and Disclaimer

Every effort has been made to make this book as complete and as accurate as possible, but no warranty or fitness is implied. The information provided is on an "as is" basis. The author and the publisher shall have neither liability nor responsibility to any person or entity with respect to any loss or damages arising from the information contained in this book.

Special Sales

For information about buying this title in bulk quantities, or for special sales opportunities (which may include electronic versions; custom cover designs; and content particular to your business, training goals, marketing focus, or branding interests), please contact our corporate sales department at corpsales@pearsoned.com or (800) 382-3419.

For government sales inquiries, please contact governmentsales@pearsoned.com.

For questions about sales outside the U.S., please contact international@pearsoned.com.

Editor-in-Chief
Greg Wiegand

Executive Editor
Rick Kughen

Development Editor
Keith Cline

Managing Editor
Kristy Hart

Project Editor
Andy Beaster

Copy Editor
Keith Cline

Indexer
Cheryl Lenser

Proofreader
Sarah Kearns

Technical Editor
Jason Foster

Publishing Coordinator
Kristen Watterson

Cover Designer
Mark Shirar

Compositor
Nonie Ratcliff

Contents at a Glance

Table of Contents

About the Author

Richard Blum has worked in the IT industry for more than 25 years as a network and systems administrator, managing Microsoft, UNIX, Linux, and Novell servers for a network with more than 3,500 users. He has developed and teaches programming and Linux courses via the Internet to colleges and universities worldwide. Rich has a master's degree in management information systems from Purdue University and is the author of several programming books, including *Teach Yourself Python Programming for the Raspberry Pi in 24 Hours* (coauthored with Christine Bresnahan, 2013, Sams Publishing), *Linux Command Line and Shell Scripting Bible* (coauthored with Christine Bresnahan, 2011, Wiley), *Professional Linux Programming* (coauthored with Jon Masters, 2007, Wiley), and *Professional Assembly Language* (2005, Wrox). When he's not busy being a computer nerd, Rich enjoys spending time with his wife, Barbara, and two daughters, Katie Jane and Jessica.

Dedication

To my Uncle George.

Thanks for all your mentoring and troubleshooting help in my early electronics projects. I never would have gotten started in my career had those projects not worked!

"Iron sharpens iron, and one man sharpens another." —Proverbs 27:17 (ESV)

Acknowledgments

First, all glory and praise go to God, who through His Son, Jesus Christ, makes all things possible and gives us the gift of eternal life.

Many thanks go to the fantastic team of people at Sams Publishing for their outstanding work on this project. Thanks to Rick Kughen, the executive editor, for offering us the opportunity to work on this book and keeping things on track, and to Andrew Beaster for all his production work. I would also like to thank Carole Jelen at Waterside Productions, Inc., for arranging this opportunity and for helping out in my writing career.

I am indebted to the technical editor, Jason Foster, who put in many long hours double-checking all the work and keeping the book technically accurate, all while getting a new job, having a new baby (congrats!), and moving to a new house in another state. His suggestions and eagle eyes have made this a much better book.

Finally I'd like to thank my wife, Barbara, and two daughters, Katie Jane and Jessica, for their patience and support while I was writing this.

We Want to Hear from You!

As the reader of this book, *you* are our most important critic and commentator. We value your opinion and want to know what we're doing right, what we could do better, what areas you'd like to see us publish in, and any other words of wisdom you're willing to pass our way.

We welcome your comments. You can email or write to let us know what you did or didn't like about this book—as well as what we can do to make our books better.

Please note that we cannot help you with technical problems related to the topic of this book.

When you write, please be sure to include this book's title and author as well as your name and email address. We will carefully review your comments and share them with the author and editors who worked on the book.

Email: consumer@samspublishing.com

Mail: Sams Publishing
ATTN: Reader Feedback
800 East 96th Street
Indianapolis, IN 46240 USA

Reader Services

Visit our website and register this book at informit.com/register for convenient access to any updates, downloads, or errata that might be available for this book.

Introduction

Since being introduced in 2005 as a student project, the Arduino microcontroller has quickly become a favorite of both hobbyists and professionals. It's a popular platform for creating many different types of automated systems—from monitoring water levels in house plants to controlling high-level robotic systems. These days you can find an Arduino behind lots of different electronic systems.

To control the Arduino, you need to know the Arduino programming language. The Arduino programming language derives from the C programming language, with some added features unique to the Arduino environment. However, beginners sometimes find the C programming somewhat tricky to navigate.

Programming the Arduino

The goal of this book is to help guide both hobbyists and students through using the Arduino programming language on an Arduino system. You don't need any programming experience to benefit from this book; I walk through all the necessary steps to get your Arduino programs up and running.

- Part I, "The Arduino Programming Environment," starts things out by walking through the core Arduino system and demonstrating the process of creating an Arduino program (called a sketch):

  Hour 1, "Introduction to the Arduino," shows the different Arduino models currently available and describes how each differs.

  Hour 2, "Creating an Arduino Programming Environment," shows how to load the Arduino IDE on a workstation and how to connect your Arduino to your workstation to get your sketches running on your Arduino.

  Hour 3, "Using the Arduino IDE," walks through all the features available to you in the IDE.

Hour 4, "Creating an Arduino Program," demonstrates the steps to build an Arduino circuit, design a sketch to control the circuit, and upload the sketch to the Arduino to run the circuit.

- Part II, "The C Programming Language," takes an in-depth look at the features of the C programming language that you need to know to write your Arduino sketches:

  Hour 5, "Learning the Basics of C," shows you how to use variables and math operators in C to manage data and implement formulas in your Arduino sketches.

  Hour 6, "Structured Commands," shows how to add logic to your sketches.

  Hour 7, "Programming Loops," demonstrates the different ways the Arduino language allows you to iterate through data, minimizing the amount of code you need to write.

  Hour 8, "Working with Strings," introduces the concept of storing and working with text values in your Arduino sketches.

  Hour 9, "Implementing Data Structures," walks through more complicated ways of handling data in sketches.

  Hour 10, "Creating Functions," provides useful tips to help minimize the amount of repeating code in your sketches.

  Hour 11, "Pointing to Data," introduces the complex topic of using pointers in the C language and shows how you can leverage their use in your sketches.

  Hour 12, "Storing Data," walks you through how to use the EEPROM storage available in the Arduino to store data between sketch runs.

  Hour 13, "Using Libraries," finishes the in-depth C language discussion by showing how to use prebuilt libraries in your sketches and how to create your own.

- Part III, "Arduino Applications," walks through the details for using your Arduino in different application environments:

  Hour 14, "Working with Digital Interfaces," shows how to read digital sensor values and use those values in your sketch and how to output digital values.

  Hour 15, "Interfacing with Analog Devices," shows how to read analog sensor values and how to use pulse width modulation to emulate an analog output voltage.

  Hour 16, "Adding Interrupts," demonstrates how to use asynchronous programming techniques in your Arduino sketches while monitoring sensors.

  Hour 17, "Communicating with Devices," covers the different communications protocols built in to the Arduino, including SPI and I2C.

Hour 18, "Using Sensors," takes a closer look at the different types of analog and digital sensors the Arduino supports and how to handle them in your sketches.

Hour 19, "Working with Motors," walks through how to control different types of motors from your Arduino sketch.

Hour 20, "Using an LCD," provides instructions on how to utilize digital displays to output data from your sketch.

Hour 21, "Working with the Ethernet Shield," discusses how to connect your Arduino to a network.

Hour 22, "Implementing Advanced Ethernet Programs," demonstrates how to provide sensor data to remote network clients and how to control the Arduino from a remote client.

Hour 23, "Handling Files," shows how to use SD card interfaces found on some Arduino shields to store data for long term.

Hour 24, "Prototyping Projects," walks you through the process of creating a complete Arduino project, from design to implementation.

Who Should Read This Book?

This book is aimed at readers interested in getting the most out of their Arduino system by writing their own Arduino sketches, including these three groups:

- Students interested in an inexpensive way to learn electronics and programming
- Hobbyists interested in monitoring and controlling digital or analog circuits
- Professionals looking for an inexpensive platform to use for application deployment

If you are reading this book, you are not necessarily new to programming, but you may be new to the Arduino environment and need a quick reference guide.

Conventions Used in This Book

To make your life easier, this book includes various features and conventions that help you get the most out of this book and out of your Arduino:

Steps—Throughout the book, I've broken many coding tasks into easy-to-follow step-by-step procedures.

Things you type—Whenever I suggest that you type something, what you type appears in a **bold** font.

Filenames, folder names, and code—These things appear in a `monospace` font.

Commands—Commands and their syntax use **bold**.

Menu commands—I use the following style for all application menu commands: *Menu, Command,* where *Menu* is the name of the menu you pull down and *Command* is the name of the command you select. Here's an example: File, Open. This means you select the File menu and then select the Open command.

This book also uses the following boxes to draw your attention to important or interesting information:

BY THE WAY

By the Way boxes present asides that give you more information about the current topic. These tidbits provide extra insights that offer better understanding of the task.

DID YOU KNOW?

Did You Know? boxes call your attention to suggestions, solutions, or shortcuts that are often hidden, undocumented, or just extra useful.

WATCH OUT!

Watch Out! boxes provide cautions or warnings about actions or mistakes that bring about data loss or other serious consequences.

PART I

The Arduino Programming Environment

HOUR 1
Introduction to the Arduino

What You'll Learn in This Hour:

- What the Arduino is all about
- The Arduino family of microcontrollers
- What programs you can run on an Arduino

In just a short amount of time, the Arduino has taken the electronics world by storm. The concept of an open source hardware platform for creating microcontroller applications has sparked an interest in both hobbyists and professionals who are looking for simple ways to control electronic projects. This hour introduces you to the Arduino microcontroller family and walks through just what the excitement is all about.

What Is an Arduino?

The Arduino is an open source microcontroller platform that you use for sensing both digital and analog input signals and for sending digital and analog output signals to control devices. That definition is quite a mouthful. Before we delve into programming the Arduino, let's take some time to look at what it's all about. This section explores just what an Arduino is and how to use it in your electronics projects.

Arduino Is a Microcontroller

These days, just about everything has a computer built in. From toasters to televisions, it's hard to find a device that doesn't have some type of computer system controlling it.

However, the computer chip embedded in your toaster differs significantly from the computer chip used in your workstation. Computer chips embedded in household devices don't need nearly as much computing power as those found in workstations. They mostly just need to monitor a few analog or digital conditions (such as the time or temperature) and be able to control a few devices (such as the heating elements in the toaster or the channel to display on the television).

Because embedded computers require less processing power, it doesn't make any sense to use the same expensive high-powered computer chips found in workstations in our toasters and televisions. Instead, developers use less-expensive computer chips that have lesser processing power in those devices.

This is where microcontrollers come into play. A microcontroller is a computer chip with minimal processing power and that often is designed for automatically controlling some type of external device. Instead of having a large feature set and lots of components squeezed into the chip, microcontrollers are designed with simplicity in mind.

Because microcontrollers don't need large processors, they have plenty of room on the chip to incorporate other features that high-powered computer chips use external circuits for. For example, most microcontrollers contain all their memory and input/output (I/O) peripheral ports on the same chip. This provides a single interface to all the features of the microcontroller, making it easier to incorporate into electronic circuits. Figure 1.1 shows a block diagram of a basic microcontroller layout.

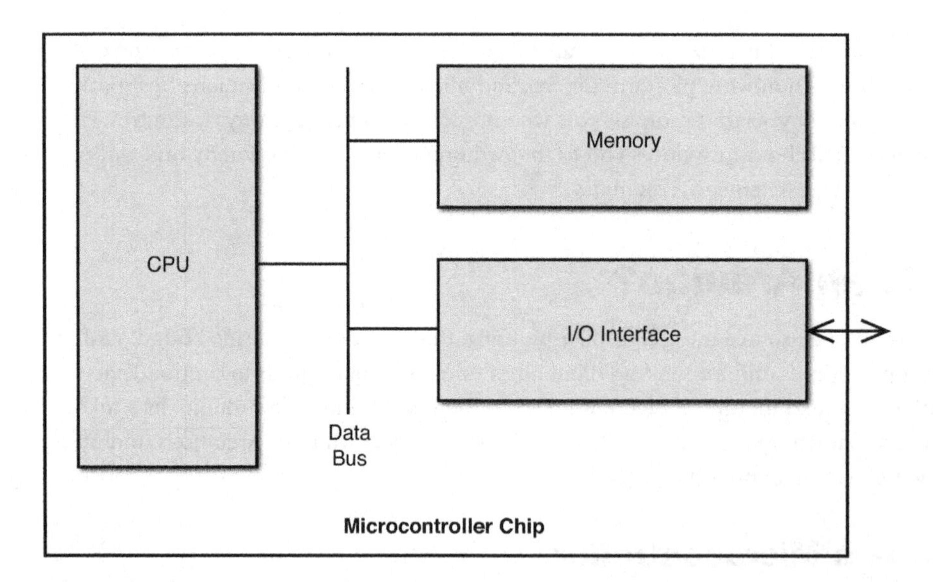

FIGURE 1.1
Block diagram of a microcontroller.

By incorporating the processor, memory, and I/O ports onto a single chip, microcontrollers provide a simple interface for designers to use when embedding them into projects.

Using Open Source Hardware

If you've heard of the Linux operating system, you're probably familiar with the idea of open source software. In the world of open source software, a group of developers release the software program code to the general public so that anyone can make suggestions for changes and bug fixes. This often results in feature-rich software that has fewer bugs.

The idea of open source hardware is much the same, except with using physical hardware rather than software. With open source hardware projects, the physical hardware that you use to control devices is open to the general public to freely use and modify as needed.

The Arduino project developers have designed a full microcontroller system that uses a standard interface to interact with external devices. The design plans and architecture have been released to the public as open source, allowing anyone both to use the Arduino free of charge in their own projects and to even modify the Arduino without violating any patent or copyright laws.

CAUTION

Using the Arduino Name

Although the Arduino hardware is open source, the Arduino name enjoys trademark protection. Anyone can create his or her own device based on the Arduino hardware, but must refrain from naming it *Arduino*. Only the Arduino project can use the Arduino name in officially released versions of the Arduino project.

The next section walks through the basics of the Arduino architecture, showing just what makes up an Arduino.

Examining the Arduino Architecture

The developers with the Arduino project have built a completely self-contained microcontroller system on a single circuit board that you can plug into just about any embedded system. The core of the Arduino project is the microcontroller chip that it uses. To be able to program the Arduino, you'll want to know a little about what's going on "under the hood."

The ATmega AVR Microcontroller Family

The Arduino developers selected the Atmel ATmega AVR family of microcontroller chips to power the Arduino. The ATmega AVR microcontroller is an 8-bit computer that contains several features built in:

- Flash memory for storing program code statements
- Static random-access memory (SRAM) for storing program data
- Erasable electronic programmable read-only memory (EEPROM) for storing long-term data

- Digital input and output ports
- Analog-to-digital converter (ADC) for converting analog input signals to digital format

The ATmega AVR chip is a complete microcontroller system built in to a single chip. That allows the Arduino developers to create a small circuit board with minimal external components.

The Arduino series uses several different ATmega processors. Table 1.1 shows the different ATmega processors you'll find in Arduino units.

TABLE 1.1 The ATmega Processor Family

Processor	Flash	SRAM	EEPROM
ATmega48	4KB	512B	256B
ATmega88	8KB	1KB	512B
ATmega168	16KB	1KB	512B
ATmega328	32KB	2KB	1KB
ATmega32u4	32KB	2.5KB	1KB

Compared to memory sizes used in workstation computers (often measured by the gigabit), the ATmega family of microcontrollers has very limited memory. However, the programs that you'll need to create for a microcontroller are considerably simpler, so most of the time you won't have to worry about the memory limitations.

The Arduino Layout

Another key to the Arduino is how it interfaces with external devices. The Arduino developers created a standard interface so that other developers could easily incorporate the Arduino directly into their projects. (This is part of the open source hardware method.)

The Arduino uses header sockets to make the input and output pins of the ATmega AVR chip available to external devices. Figure 1.2 shows the layout of the Arduino Uno.

You can directly access all the microcontroller interfaces from the header sockets. The bottom set of sockets contain the analog input interfaces and access to the voltage and ground pins on the microcontroller. The top header socket contains the digital I/O interfaces. The standard layout of the header sockets allows developers to easily build plug-in devices (called shields) that interface with the Arduino.

Now that you've seen the basics of the Arduino, the next section takes a closer look at the different Arduino units available. There are plenty of different Arduino units to choose from for use in your projects, so knowing which one to use can sometimes be confusing.

FIGURE 1.2
The Arduino Uno layout.

Introducing the Arduino Family

If you've started looking for an Arduino unit to work with, you've probably noticed that there's not just one Arduino. This section covers the history of the Arduino unit and why so many different versions of it are available.

The History of the Arduino

Interestingly enough, the Arduino wasn't designed by a large electronics corporation or even by a group of computer science majors. Instead, it was designed out of necessity by a group of students and instructors looking for a solution to animate their art projects.

Modern art projects often require synchronized moving parts, which necessitates precise automation, which in turn, requires some type of microcontroller system.

Because most art students aren't by nature programmers, having to purchase microcontroller chips and design electronic circuits to make them run became quite a challenge. A group of students at the Interaction Design Institute Ivera (IDII) in Italy worked on a project to help minimize the amount of coding art students had to write to automate their artistic creations.

This resulted in the Wiring project, which produced a standard microcontroller circuit board, along with a standard programming environment, called Processing. Created in 2003, the Wiring and Processing projects gained some acceptance, but were somewhat expensive for most students to experiment with.

After a few years of tweaking the Wiring project, in 2005 a group of designers led by Massimo Banzi and David Cuartielles came up with the Arduino project. The Arduino project built upon the basic features of the Wiring project, but at a lower cost for students.

Every part of the Arduino system was designed for simplicity for nontechnical people. The hardware interface is somewhat forgiving. For instance, if you hook up the wrong wires to the wrong ports, you won't usually blow up your Arduino unit. If you do happen to manage to blow up the ATmega microcontroller chip, the Arduino was designed to easily replace the microcontroller chip without having to purchase an entire Arduino unit.

Likewise, the software for the Arduino was designed with nonprogrammers in mind. In an interesting tie to the art world, programs that you create for the Arduino are called sketches, and the folders where you store your sketches are called sketchbooks.

Exploring the Arduino Models

Part of the open source hardware method is to provide lots of options for developers. This allows developers to find just the right Arduino to fit into their project. The following sections walk through the different Arduino units currently available.

Arduino Uno

The workhorse of the Arduino family is the Uno. It's the most popular Arduino unit and provides the basic functionality of the ATmega AVR microcontroller within the standard Arduino footprint.

It uses an ATmega328 microcontroller, which provides 14 digital I/O interfaces, 6 analog input interfaces, and 6 pulse-width modulation (PWM) interfaces for controlling motors.

The Uno circuitry also contains a USB interface for communicating with workstations as a serial communications device, a separate power jack so that you can power the Arduino Uno without plugging it into a workstation, and a reset button to restart the program running on the Arduino.

The size and layout of the Arduino Uno has been made the standard format for most Arduino units, so just about all Arduino shields (discussed later in the "Exploring Arduino Shields" section) fit into the header sockets of the Arduino Uno.

TIP

Following Along with the Book

All the examples in this book use the Arduino Uno unit. If you want to follow along with the examples, ideally you should use the Arduino Uno, although the other full-sized Arduino units would work too, but with some modifications to the projects.

Arduino Due

The Arduino Due unit uses a more powerful 32-bit ARM Cortex-M3 CPU instead of the standard ATmega328 microcontroller. This provides significantly more processing power than the standard Arduino Uno.

The Due provides 54 digital I/O interfaces, 12 PWM outputs, 12 analog inputs, and 4 serial Universal Asynchronous Receiver/Transmitter (UART) interfaces. It also provides considerably more memory, with 96KB of SRAM for data storage, and 512KB of flash memory for program code storage.

The Arduino Due uses the standard Uno header layout, but adds another header along the right side of the unit, providing the extra interfaces. This allows you to use most shields designed for the Uno on the Due.

Arduino Leonardo

The Arduino Leonardo uses the ATmega32u4 microcontroller, which provides 20 digital I/O interfaces, 7 PWM outputs, and 12 analog inputs. The Arduino Leonardo has the same header layout as the Uno, with additional header sockets for the extra interfaces.

One nice feature of the Arduino Leonardo is that it can emulate a keyboard and mouse when connected to a workstation using the USB port. You can write code to run on the Arduino Leonardo that sends standard keyboard or mouse signals to the workstation for processing.

Arduino Mega

The Arduino Mega provides more interfaces than the standard Arduino Uno. It uses an ATmega2560 microcontroller, which provides 54 digital I/O interfaces, 15 PWM outputs, 16 analog inputs, and 4 UART serial ports. You can use all the standard Arduino shields with the Mega unit.

The Arduino Mega provides a lot more interfaces, but it comes at a cost of a larger footprint. The Arduino Uno is a small unit that can easily fit into a project, but the Arduino Mega device is considerably larger.

Arduino Micro

The Arduino Micro is a small-sized Arduino unit that provides basic microcontroller capabilities, but in a much smaller footprint. It uses the same ATmega32u4 microcontroller as the Lenardo, which provides 20 digital I/O interfaces, 7 PWM outputs, and 12 analog inputs.

The selling point of the Arduino Micro is that it is only 4.8cm long and 1.77cm wide, small enough to fit into just about any electronics project. It uses a USB port to communicate as a serial device with workstations, or as a keyboard and mouse emulator like the Lenardo can.

The downside to the Micro is that because of its smaller size, it doesn't work with the standard Arduino shields, so you can't expand its capabilities with other features.

Arduino Esplora

The Arduino Esplora is an attempt at creating an open source game controller. It uses the ATmega 32u4 microcontroller just like the Lenardo and Micro do, but also contains hardware commonly found on commercial game controllers:

- Analog joystick
- Three-axis accelerator
- Light sensor
- Temperature sensor
- Microphone
- Linear potentiometer
- Connector for an LCD display
- Buzzer
- LED lights
- Switches for up, down, left, and right

The Esplora also has a unique design to help it fit into a game controller case that could be handheld, providing easier access to the switches and analog joystick.

Arduino Yun

The Arduino Yun is an interesting experiment in combining the hardware versatility of an ATMega32u4 microcontroller with an Atheros microprocessor running a Linux operating system environment. The two chips communicate with each other directly on the Yun circuit board, so that the ATmega microcontroller can pass data directly to the Atheros system for processing on the Linux system.

The Linux system also includes built-in Ethernet and Wi-Fi ports for external communication, in addition to a standard USB port to run a remote serial console. This is the ultimate in embedding a full Linux system with a microcontroller.

Arduino Ethernet

The Arduino Ethernet project combines the microcontroller of the Arduino Uno with a wired Ethernet interface in one unit. This enables you to interface with the Arduino remotely across a network to read data collected or to trigger events in the microcontroller.

LilyPad Arduino

The Arduino LilyPad is certainly an interesting device. It was designed for the textile arts world, embedding a microcontroller within textiles. Not only is it small, but it's also very thin and lightweight, perfect for sewing within just about any type of fabric. Think of wearing a shirt with built-in LEDs that flash messages to your friends!

One interesting feature of the LilyPad Arduino is the built-in MCP73831 LiPo battery charging chip. This chip allows you to embed rechargeable batteries with the device; when the unit is plugged into a USB port of a workstation, the rechargeable batteries recharge *automatically*.

Examining the Arduino Uno

At the time of this writing, the Arduino Uno R3 is the current main board in the Arduino family. It's the most commonly used unit for interacting with projects that require a simple microcontroller. The Uno R3 unit provides the following:

- 14 digital I/O interfaces
- 6 analog input interfaces
- 6 PWM interfaces
- 1 I2C controller interface
- 1 SPI controller interface
- 1 UART serial interface, connected to a USB interface

The Arduino Uno circuit board, shown back in Figure 1.2, uses the standard Arduino header socket layout.

The bottom row of sockets contains the analog input sockets on the right, along with sockets that provide power for external circuits on the left. Along the top of the Arduino are the digital I/O sockets.

The digital I/O sockets have a double use. Not only are they used for digital input or output signals, but some of them are used for secondary purposes, depending on how you program the ATmega microcontroller:

- Digital sockets 0 and 1 are also used as the RX and TX lines for the serial UART.
- Digital sockets 3, 5, 6, 9, 10, and 11 are used for PWM outputs.
- The leftmost four sockets in the top header socket row are for the SPI controller interface.

Besides the header sockets, the Arduino Uno provides a standard USB port for connecting the unit to a workstation. The USB port uses the UART serial interface of the ATmega microcontroller to send data to the microcontroller. This is how you load programs into the Arduino. You can also use the USB serial interface to view output from the microcontroller, which comes in handy when debugging your Arduino sketches.

The Arduino Uno also has four built-in LEDs, shown in Figure 1.3, that help you see what's going on in the Arduino:

- A green LED (marked ON) that indicates when the Arduino is receiving power
- Two yellow LEDs (marked TX and RX) that indicate when the UART serial interface is receiving or sending data
- A yellow LED (marked L) connected to digital output socket 13 that you can control from your programs

Finally, one new addition to the R3 version of the Arduino Uno is a Reset button, located at the upper-left corner of the circuit board. Pressing the Reset button forces the ATmega microcontroller to reboot, which reloads the program code in memory and starts executing code from the beginning of the program.

Accessories You Might Need

Besides picking up an Arduino unit, you'll also need to pick up a few other components to complete your project. This section identifies the more common parts you'll need:

- **A USB A-B cable:** The Arduino USB port uses a B-type USB interface, so you'll need to find an A-B USB cable to connect the Arduino to your workstation. Most printers use a B-type USB interface, so these cables aren't hard to find.
- **An external power source:** After you program your Arduino, you probably don't want to keep it tethered to your workstation. The Arduino Uno includes a 2.1mm center positive power jack so that you can plug in an external power source. This can either be an AC/DCV converter or a batter power pack that provides 5V power.

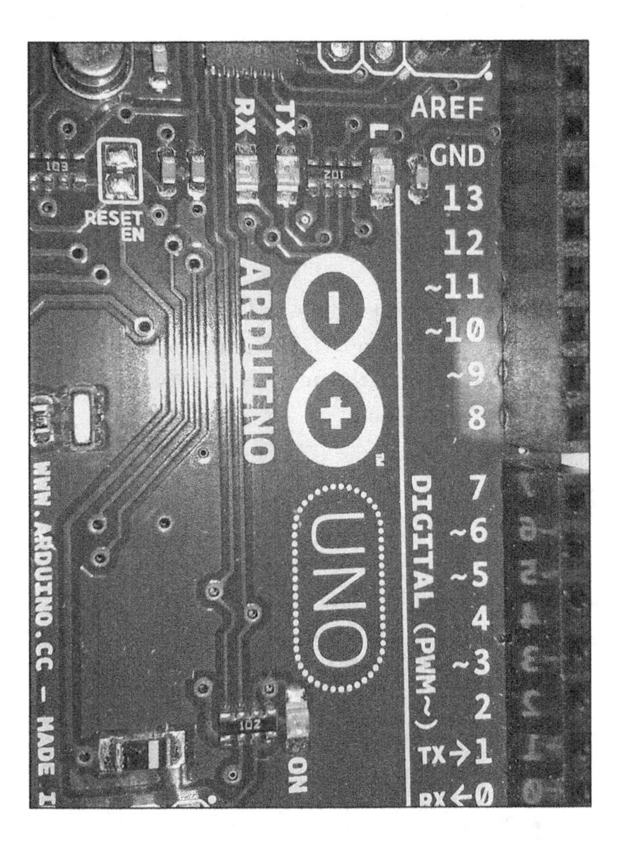

FIGURE 1.3
The Arduino Uno R3 LEDs.

- **A breadboard:** As you experiment with interfacing various electronic components to your Arduino, a breadboard comes in handy to quickly build connections. Once you have your circuits designed, you can purchase prototype shields to make the interface connections more permanent (see Hour 24, "Prototyping Projects").
- **Resistors:** Resistors are used for limiting the current through a circuit, usually to avoid burning out the microcontroller inputs or the LED outputs.
- **Variable resistors:** You can also use variable resistors (called potentiometers) to adjust the voltage going into an analog input socket to control a circuit.
- **Switches:** Switches allow you to provide a digital input for controlling your programs.
- **Wires:** You'll need some wires for connecting the breadboard components together and interfacing them into the Arduino header sockets.

- **Sensors:** You can interface a number of both digital and analog sensors into your Arduino. There are temperature sensors and light sensors, for instance, that produce varying analog outputs based on the temperature or amount of light.
- **Motors:** Controlling moving parts requires some type of motor. The two main types of motors you'll most often require are servo motors, which rotate to a fixed position, and DC motors, which can spin at a variable rate based on the voltage provided.

Often you can find different Arduino kits that include these parts. The Arduino Store provides an official Arduino Starter Kit package, which includes an Arduino Uno R3 unit, a USB cable, a breadboard, and parts to create 15 separate projects (along with the tutorials on how to build them). This kit is a great way to get started with your Arduino unit.

Exploring Arduino Shields

The beauty of using open source hardware for your project is that plenty of other developers have already solved many of the same issues that you'll run into and are willing to share their solutions. The standard interface to the Arduino provides a common way for developers to build external circuits that interface to the Arduino and thus provide additional functionality.

These units are called shields. There are plenty of Arduino shields available for a myriad of different projects. This section discusses a few popular shields.

Connecting with the Ethernet Shield

These days just about every project needs some type of network connectivity. Microcontroller projects can use a network to send data to a remote monitoring device, to allow remote connectivity to monitor data, or to just store data in a remote location.

The Ethernet shield provides a common wired network interface to the Arduino, along with a software library for using the network features. Figure 1.4 shows the Ethernet shield plugged into an Arduino Uno R3 unit.

While the Ethernet shield plugs into the standard header sockets of the Arduino Uno, it also has the same standard header sockets built in, so you can piggy-back additional shields onto the Ethernet shield (so long as they don't use the same signals to process data).

Displaying Data with the LCD Shield

Often, instead of connecting to the Arduino unit to retrieve information, it would be nice to just view data quickly. This is where the LCD shield comes into play.

FIGURE 1.4
The Ethernet shield plugged into the Arduino Uno.

It uses a standard LCD display to display 2 rows of 25 characters that you can easily view as the Arduino unit is running. The LCD shield also contains five push buttons, enabling you to control the program with basic commands.

Running Motors with the Motor Shield

A popular use for the Arduino is to control motors. Motors come in handy when working with robotics and when making your project mobile using wheels.

Most motors require the use of PWM inputs so that you can control the speed of the motor. The basic Arduino system contains six PWM outputs, enabling you to control up to six separate motors. The motor shield expands that capability.

Developing New Projects with the Prototype Shield

If you plan on doing your own electronic circuit development, you'll need some type of environment to build your circuits as you experiment. The Prototype shield provides such an environment.

The top of the Prototype shield provides access to all the Arduino header pins, so you can connect to any pin in the ATmega microcontroller. The middle of the Prototype shield provides standard-spaced soldering holes for connecting electronic parts, such as resistors, sensors, diodes, and LEDs.

If you would like to experiment with your projects before soldering the components in place, you can also use a small solderless breadboard on top of the Prototype shield, which allows you to create temporary layouts of your electronic components for testing, before committing them to the soldering holes.

Summary

This hour covered the basics of what the Arduino is and how you can use it in your electronic projects. You first learned what microcontrollers are and how you can use them to sense digital and analog inputs and to output digital signals to control devices. The discussion then turned to the Arduino and how it uses the ATmega AVR microcontroller along with a standard interface to receive analog and digital signals and to send digital output signals. You then learned about the different Arduino shields available for adding functionality to your Arduino projects, such as connecting it to a network or using it to control motors and servos.

In the next hour, we take a closer look at what you need to start programming the Arduino.

Workshop

Quiz

1. Which type of memory should you use to store data that you can retrieve after the Arduino has been powered off?
   - **A.** Flash memory
   - **B.** SRAM memory
   - **C.** EEPROM memory
   - **D.** ROM memory

2. The Arduino requires a connection to a workstation to work. True or false?

3. Is Arduino shield available that allows your program to display simple information without requiring a connection to a computer?

Answers

1. C. The EEPROM memory is the only data memory built in to the Arduino that can retain data when the unit is powered off. The flash memory can also retain information after a power off, but it's used only to store program code, not data.

2. False. All the Arduino models allow you to connect a battery or AC/DC converter to run the Arduino without it being plugged into a computer.

3. Yes, the LCD shield provides a 2-row, 25-character display that your Arduino programs can use to display simple data while running.

Q&A

Q. I'm new to electronics and don't know what parts to get to work with my Arduino. Are there complete kits available that include the electronics required to run projects?

A. Yes, the Arduino project has a complete kit available that includes an Arduino Uno along with motors, an LCD display, sensors, and a breadboard and parts required to build 15 projects. You can find the kit, along with other Arduino units and shields, at various online retailers, such as AdaFruit, SparkFun, and Newark Electronics.

Q. Do you have to reprogram the Arduino each time you run it?

A. No, the Arduino stores the program in Flash memory, which retains the program code after a power off. You only have to load your program once, and the Arduino will retain it no matter how often you turn it on or off.

HOUR 2

Creating an Arduino Programming Environment

What You'll Learn in This Hour:

- How microcontrollers work
- How microcontroller programs work
- What the Arduino Programming Language is
- How to download and install the Arduino IDE

Just having an Arduino hardware platform isn't enough to get your projects working. Besides the hardware, you'll also need to write a program to tell the Arduino hardware what to do. This hour walks through the basics of how microcontrollers work, how they read and run programs, and how to set up an environment to create programs for your Arduino.

Exploring Microcontroller Internals

Before you can start programming the Arduino, it helps to know a little bit about what's going on inside of it. At the core of all Arduino units is an ATmega AVR series microcontroller. The programs that you write to control the Arduino must run on the ATmega microcontroller. This section examines the layout of the ATmega microcontroller, showing you the different components and how they interact to run your programs.

Peeking Inside the ATmega Microcontroller

To refresh your memory from the first hour, three basic components reside inside of a microcontroller:

- The CPU
- Memory
- The input/output (I/O) interface

Each of these components plays a different role in how the microcontroller interfaces with your project and in how you program the microcontroller. Figure 2.1 shows the layout of these components in a microcontroller.

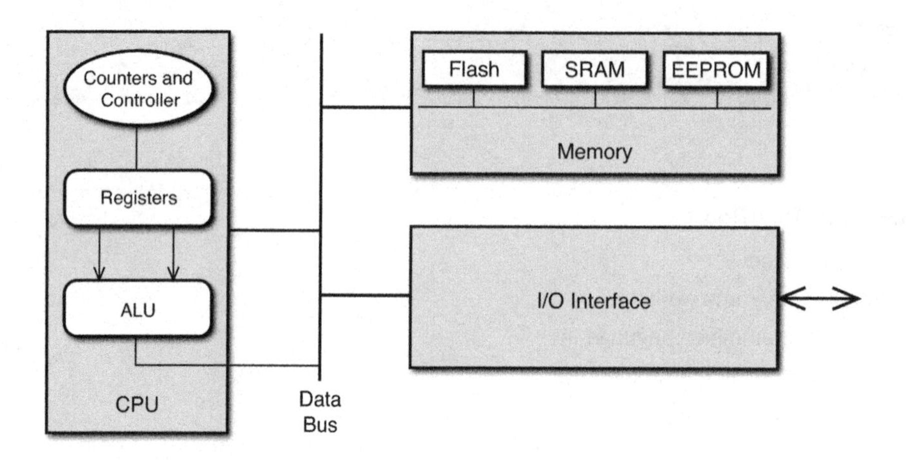

FIGURE 2.1
The ATmega microcontroller block layout.

Now let's take some time to look inside each of these individual components and see how they work together to run your programs.

The CPU

The CPU inside the microcontroller isn't all that different from the CPU you would find in your workstation computer, just a little smaller. It consists of several common parts found in most computer systems:

- An Arithmetic Logic Unit (ALU), which performs the mathematical and logical operations
- Data registers, which store data for processing in the ALU
- A status register, which contains the status information of the most recently executed arithmetic instruction
- A program counter, which keeps track of the memory location of the next instruction to run
- A stack pointer, which keeps track of the location of temporary data stored in memory
- A controller, which handles loading program code and memory data into the registers, ALU, and pointers

All the CPU components interact within the CPU based on a clocking signal provided to the microcontroller. At each clock pulse, the CPU performs one operation. The speed of the clock pulse determines how fast the CPU processes instructions.

For example, the Arduino Uno utilizes a 16MHz clock, which provides 16 million clock pulses per second. Most of the ATmega microcontroller CPU instructions require only one or two clock pulses to complete, making the Arduino capable of sampling inputs at a very high rate.

Memory

The Atmel ATmega microcontroller family uses what's called the Harvard architecture for computers. This model separates the computer memory into two components: one memory section to store the program code, and another memory section to store the program data.

The ATmega microcontroller family uses a separate form of memory for each type:

- Flash memory to store program code
- Static random-access memory (SRAM) to store program data

The flash memory retains data after you remove power from the microcontroller. That way your program code stays in memory when you power off the Arduino. When you turn the Arduino back on, it starts running your program immediately without you having to reload it each time.

SRAM memory does not retain data when the power is off, so any data your program stores in SRAM will be lost at power down. To compensate for that, the ATmega microcontroller also incorporates a separate electronically erasable programmable read-only memory (EEPROM) memory section besides the SRAM for storing data.

The EEPROM memory can retain data after loss of power, until you manually clear it out. It's not as fast as the SRAM memory, so you should use it only when you need to save data for later use.

Input/Output interface

The I/O interface allows the microcontroller to interact with the outside world. This is what allows the ATmega controller to read analog and digital input signals and send digital output signals back to external devices.

The ATmega controller contains a built-in analog-to-digital converter (ADC), which converts analog input signals into a digital value that you can read from your programs. It also contains circuitry to control the digital interface pins so that you can use the same pins for either digital input or digital output.

TIP

Analog Output

You might have noticed that the ATmega microcontroller doesn't have any analog output interfaces. The reason for that is you can simulate an analog output by using digital outputs and a technique called pulse-width modulation (PWM). PWM sends a digital signal of a varying duration and frequency, thus emulating an analog output signal. The ATmega328 microcontroller used in the Arduino Uno supports six PWM outputs.

All the microcontroller components work together to process input signals and generate output signals. The CPU is what controls all the action. The next section takes a closer look at just how the CPU functions in the microcontroller.

Programming the Microcontroller

You control what happens in the CPU using software program code. The CPU reads the program code instructions, and then performs the specified actions one step at a time. Each type of CPU has its own specific instructions that it understands, called its instruction set.

The instruction set is what tells the CPU what data to retrieve and how to process that data. This is where your programming comes in. Operations such as reading data, performing a mathematical operation, or outputting data are defined in the instruction set.

Because the CPU can only read and write binary data, the instruction set consists of groups of binary data, usually consisting of 1- or 2-byte commands. These commands are called machine code (also commonly called operation codes, or opcodes for short). Each instruction has a specific binary code to indicate the instruction, along with what data the instruction requires.

The ATmega AVR instruction set uses 16-bit (2-byte) instructions to control the CPU actions. For example, the instruction to add the data values stored in registers R1 and R2, and then place the result in register R1, looks like this:

```
0000 1100 0001 0010
```

When the CPU processes this instruction, it knows to fetch the data values from the registers, use the ALU to add them, and then place the result back in register R1.

Knowing just what machine codes do what functions, and require what data, can be an almost impossible task for programmers. Fortunately, there's some help with that, as discussed in the next section.

Moving Beyond Machine Code

Machine code is used directly by the microcontroller to process data, but it's not the easiest programming language for humans to read and understand. To solve that problem, developers have created some other ways to create program code. This section looks at those methods and explains how they can help with your Arduino programming.

Coding with Assembly Language

Assembly language is closely related to machine code; it assigns a text mnemonic code to represent each individual machine code instruction. The text mnemonic is usually a short string that symbolizes the function performed (such as `ADD` to add two values). That makes writing basic microcontroller programs a lot easier.

For example, if you remember from the preceding section, the machine code to add the values stored in register R1 to the value stored in register R2 and place the result in register R1 looks like this:

```
0000 1100 0001 0010
```

However, the assembly language code for that function looks like this:

```
add r1, r2
```

Now that makes a lot more sense for humans, and is a lot easier to remember for coding. Each of the machine code instructions for the ATmega AVR family of microcontrollers has an associated assembly language mnemonic. These are listed in the *AVR Instruction Set Manual*, available for download from the Atmel website.

TIP

The AVR Instruction Set

If you're curious, you can download the complete AVR instruction set from the Atmel website at http://www.atmel.com/images/doc0856.pdf.

The ATmega AVR series of microcontrollers (which is what the Arduino uses) has 282 separate instructions that they recognize, so there are 282 separate assembly language mnemonics. The instructions can be broken down into six categories:

- Load data from a memory location into a register.
- Perform mathematical operations on register values.
- Compare two register values.

- Copy a register value to memory.
- Branch to another program code location.
- Interact with a digital I/O interface.

There are separate instructions for each register, each mathematical operation, and each method for copying data to and from memory. It's not hard to see why there are 282 separate instructions.

While the Atmel AVR assembly language is an improvement over coding machine language, creating fancy programs for the Arduino using assembly language is still somewhat difficult and usually left to the more advanced programmers. Fortunately for us, there's an even easier way to program the Arduino unit, as covered in the next section.

Making Life Easier with C

Even when using the Atmel AVR assembly language, it can still be quite the challenge trying to remember just which of the 282 separate instructions you need to use for any given operation. Fortunately for us, yet another level of programming makes life much easier for programmers.

Higher-level programming languages help separate out having to know the inner workings of the microcontroller from the programming code. Higher-level programming languages hide most of the internal parts of the CPU operation from the programmer, so that the programmer can concentrate on just coding.

For example, with higher-level programming languages, you can just assign variable names for data, and the compiler converts those variable names into the proper memory locations or registers for use with the CPU. With a higher-level programming language, to add two numbers and place the result back into the same location, you just write something like this:

```
value1 = value1 + value2;
```

Now that's even better than the machine code version.

The key to using higher-level programming languages is that there must be a compiler that can convert the user-friendly program code into the machine language code that runs in the microcontroller. That's done with a combination of a compiler and a code library.

The compiler reads the higher-level programming language code and converts it into the machine language code. Because it must generate machine language code, the compile is specific to the underlying CPU that the program will run. Different CPUs require different compilers.

Besides being able to convert the higher-level programming language code into machine code, most compilers also have a set of common functions that make it easier for the programmer to write code specific for the CPU. This is called a code library. Again, different CPUs have different libraries to access their features.

Many different higher-level programming languages are available these days, but the Atmel developers chose the popular C programming language as the language to use for creating code for the AVR microcontroller family. They released a compiler for converting C code into the AVR microcontroller machine language code, in addition to a library of functions for interacting with the specific digital and analog ports on the microcontroller.

When the Arduino project developers chose the ATmega microcontroller for the project, they also wanted to make coding projects easy for nonprogrammers. To do that, they built on to the existing Atmel C language compiler and library and created yet another library of code specific to using the AVR microcontroller in the Arduino environment. The next section explores both the Atmel and Arduino C language libraries.

Creating Arduino Programs

Currently, two main C language libraries are available for the Arduino environment:

- The Atmel C library
- The Arduino project library

The Atmel C library is created for using ATmega microcontrollers in general, and the Arduino project library is created specifically for nonprogrammers to use for the Arduino unit. This section takes a brief look at the two separate Arduino programming environments to familiarize you with features each provides for the Arduino programmer.

Exploring the Atmel C Library

The Atmel C environment consists of two basic packages:

- A command-line compiler environment
- A graphical front end to the compiler environment

The AVR Libc project (http://www.nongnu.org/avr-libc/) has worked on creating a complete C language command-line compiler and library for the ATmega AVR family of microcontrollers. It consists of three core packages:

- The avr-gcc compiler for creating the microcontroller machine code from C language code
- The avr-libc package for providing C language libraries for the AVR microcontroller
- The avr-binutils package for providing extra utilities for working with the C language programs

The avr-gcc package uses the open source GNU C compiler (called gcc), and customizes it to output the AVR machine code instruction set code used in the ATmega microcontrollers. This is what allows you to write C programs that are converted into the machine code that runs on the microcontroller.

TIP

The avr-gcc Package

If you're curious, you can find a Windows version of the avr-gcc command-line compiler in the WinAVR package at http://winavr.sourceforge.net/.

However, these days just about every application uses a graphical interface, including compilers. The Atmel developers have released a full integrated development environment (IDE) package that provides a graphical windows environment for editing and compiling C programs for the AVR microcontroller.

The Atmel Studio package combines a graphical text editor for entering code with all the word processing features you're familiar with (such as cutting and pasting text) with the avr-gcc compiler. It also incorporates a full C language debugger, making it easier to debug your program code. In addition, it outputs all the error messages generated by the microcontroller into the graphical window.

Although the Atmel C library makes programming the ATmega series of microcontrollers easier, it's still considered to be a tool for advanced programmers. Instead, most Arduino beginners (and even many advanced users) use the Arduino programming tools created by the Arduino project. The next section walks through these.

TIP

Exploring the Atmel Studio Package

If you're interested in exploring the Atmel Studio software, you can find more information, including a link to download the Atmel Studio software, at http://www.atmel.com/tools/ATMELSTUDIO.aspx.

Using the Arduino Programming Tools

One downside to the Atmel C library is that it uses generic code for interfacing with the ATmega microcontroller. The benefit of the Arduino hardware is that it makes accessing the specific features of the ATmega microcontroller easier, but that's lost if you just use the Atmel C library code.

Fortunately, the Arduino development team has helped solve that problem for us by creating the Arduino Programming Language.

Examining the Arduino Programming Language

The Arduino developers have created a C library that contains additional functions to help make interacting with the Arduino features much easier than coding with the Atmel C library.

For example, to send a signal to a digital output line, you simply use the function `digitalWrite()`. To read a signal from an analog input line, you use the function `analogRead()`. That makes writing code for the Arduino much easier.

The Arduino developers released this customized C library along with the Atmel avr-gcc C language compiler in a single package, called the Arduino IDE. The Arduino IDE is a graphical interface, similar to the Atmel Studio package but not quite as complex. Figure 2.2 shows the basic Arduino IDE window.

FIGURE 2.2
The Arduino IDE software package.

The IDE includes a full editor for creating your Arduino programs, a compiler to build the finished program, and a method to upload your completed program into the Arduino unit. Creating programs to run on the Arduino is a breeze with the Arduino IDE.

The Arduino Shield Libraries

Besides the core Arduino C libraries, the Arduino developers have created libraries for all the common Arduino shield devices. This allows you to easily incorporate the features of a shield into your programs without having to write complicated code.

Specialty libraries are available for all the popular Arduino shields, such as the following:

- Ethernet
- LCD display
- Motor controller
- SD card

The Arduino IDE contains all the popular shield libraries by default, so you don't have to go hunting on the Internet looking for files to download. With the Arduino IDE and the shield libraries, you can create programs for just about any Arduino project that you'll work on.

Installing the Arduino IDE

Unfortunately, the Arduino units don't come with the Arduino IDE software. Before you can get started with the Arduino IDE, you have to download it from the Arduino website and install it on to your workstation. This section walks through that process for each of the platforms supported by the Arduino IDE.

Downloading the Arduino IDE

The Arduino project maintains a web page specifically to support the Arduino IDE package (http://arduino.cc/en/main/software). When you go to that site, you'll find links to the latest version of the Arduino IDE for three platforms:

- Windows
- OS X
- Linux

Besides packages built for these three platforms, the Arduino project also provides the source code for the complete Arduino IDE package. So, if you're adventurous (and experienced in compiling programs), you can download the source code and compile it on whatever platform you need.

CAUTION

Beta Software

The Arduino download web page often contains a link to a development version of the Arduino IDE software. This is considered beta software and therefore may contain bugs and features that don't work yet. For beginners, it's usually a good idea to avoid testing beta software until you've become more familiar with how the Arduino and the Arduino IDE work.

Just click the Download link for the package you need for your development platform. After you download the Arduino IDE package, you're ready to install it, as covered in the following sections for the three different environments.

Installing for Windows

You might have noticed that there are two download versions for the Windows platform:

- A standard Windows install file
- A zip file

The install file version contains both the Arduino IDE software and the driver files necessary for Windows to recognize the Arduino when you connect it via a USB cable. The installer runs a wizard, which automatically installs the software and drivers. It is the easiest way to install the Arduino IDE environment, and is recommended.

The zip file just contains the Arduino IDE files along with the Windows driver files. You'll need to install the driver files manually using the Windows Device Manager tool.

This section walks through the installation process for both methods.

Using the Windows Installer File

When you run the Windows installer file, a wizard utility starts up, guiding you through the software installation, as shown in Figure 2.3.

The wizard guides you through the steps for installing the Arduino IDE software. You can accept the default folder location to install the Arduino software or change it if you prefer to store the program files in another location.

After the software installs, the wizard installs the Arduino software drivers necessary for Windows to interface with your Arduino.

Just follow the prompts in the wizard to complete the software and driver installation.

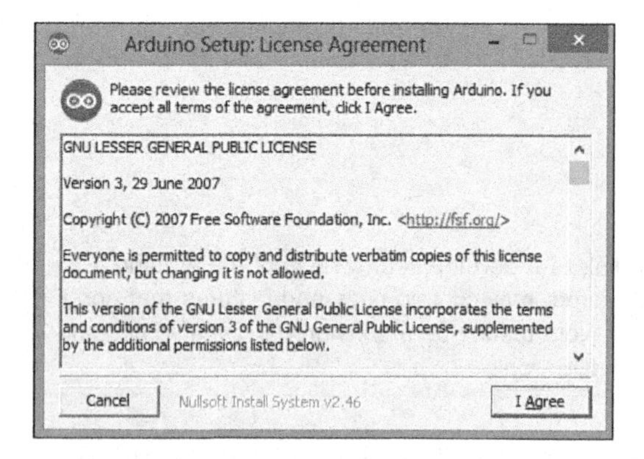

FIGURE 2.3
The Windows Arduino IDE installation wizard.

Using the Windows Zip File

If you choose to download the Windows zip file version of the software, you'll find the installation a bit more complicated. The zip file contains all the same files as the Windows installer version, but it doesn't automatically install them.

To run the Arduino IDE, just extract the zip file contents into a folder and look for the `arduino.exe` program file. You can create a shortcut to this file on your desktop for easy access.

However, before you can use the Arduino IDE, you must manually install the Windows drivers for the Arduino unit. The following steps walk through how to do that.

▼ TRY IT YOURSELF

Installing the Arduino Drivers Manually

1. After unzipping the Windows zip file to a folder, plug your Arduino unit into the USB port on your workstation.
2. The Windows USB utility will appear, but will complain that it is unable to find the driver for the Arduino unit.
3. Open the Device Manager interface in your Windows workstation, as shown in Figure 2.4.
4. Double-click the Unknown device entry.
5. Click the Update Driver button in the Unknown Device Properties dialog box.
6. Select the Browse My Computer for Driver Software option.

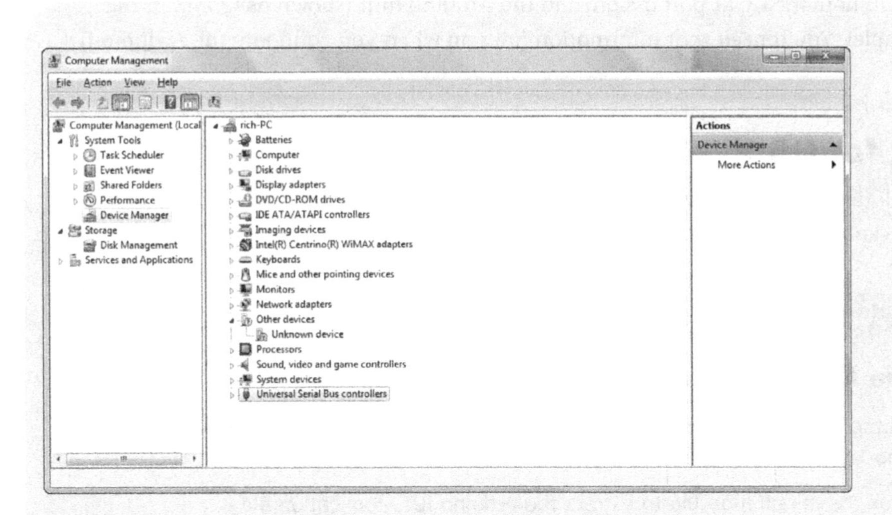

FIGURE 2.4
The Windows Device Manager showing an unknown device.

7. Navigate to the folder where you unzipped the Arduino Windows zip file in the \drivers folder.

8. Click the Install button for the Arduino USB driver dialog box.

9. The Device Manager should now show the Arduino unit listed as a COM port on your workstation, as shown in Figure 2.5.

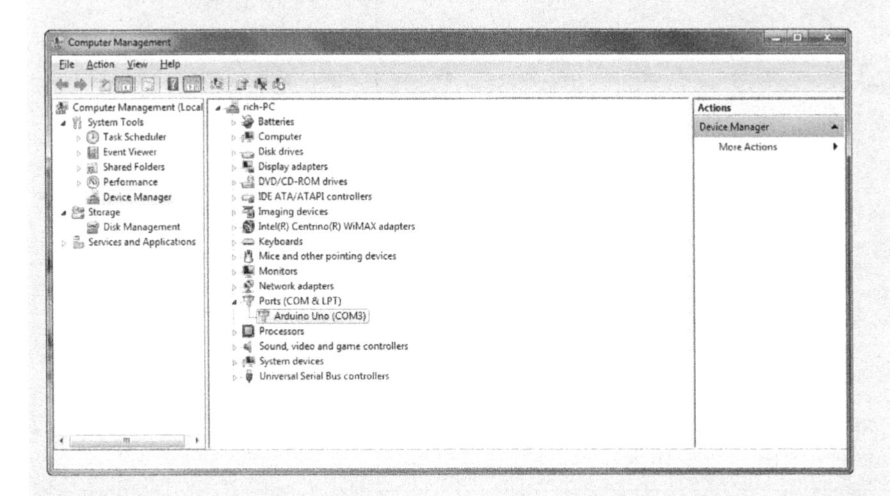

FIGURE 2.5
The installed Arduino USB driver COM port.

It's a good idea to note the COM port assigned to the Arduino unit (shown as COM3 in the Figure 2.5 example). You'll need that information later on when you configure the Arduino IDE software.

Installing for OS X

The OS X installation software provides only a zip file to download. Follow these steps to get the Arduino IDE package installed in your OS X environment.

▼ TRY IT YOURSELF

Installing the Arduino IDE for OS X

1. Download the most recent release of the OS X package for the Arduino IDE (currently the file arduino-1.0.5-macosx.zip).
2. Double-click the installation file to extract the Arduino IDE application file.
3. Move the Arduino application file to the Applications folder on your Mac.

After you install the Arduino IDE software, you can connect your Arduino using a USB cable. The OS X system will automatically recognize the Arduino as a new network interface, as shown in Figure 2.6.

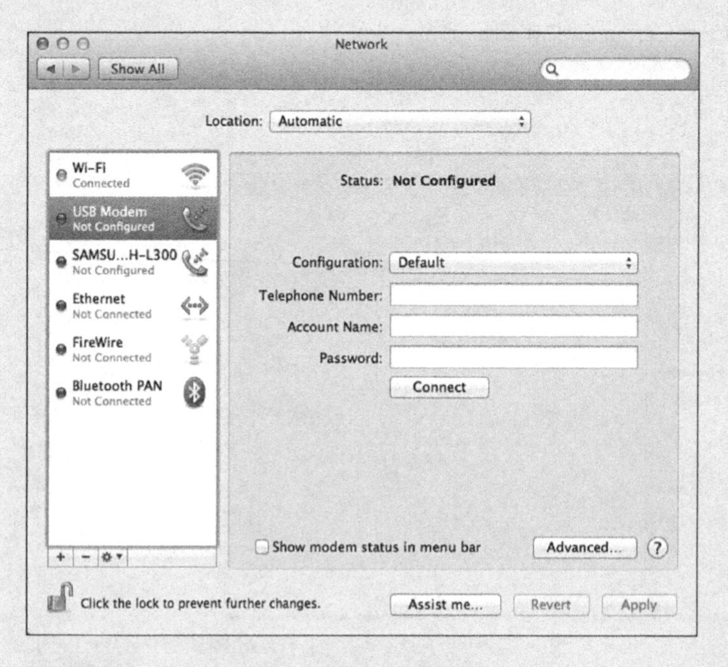

FIGURE 2.6
An OS X system detecting an Arduino device.

CAUTION

Arduino Status

After you install the Arduino device in OS X, the device will appear in a Not Configured status. Don't worry, the Arduino will work just fine.

Installing for Linux

The Linux installation provides for two separate zip files:

- A 32-bit version
- A 64-bit version

You'll need to select the version appropriate for your specific Linux distribution. Both versions are distributed as a `.tar.gz` file format, commonly called a tarball in Linux circles.

To extract the tarball into a folder, you use the `tar` command, and add the `-z` option to uncompress the tarball as it extracts the files:

```
rich@myhost ~> tar -zxf arduino.tar.gz
```

The resulting file is the Arduino IDE executable program file. Just create a shortcut on your desktop to the file.

TIP

Arduino and Ubuntu

The Ubuntu Linux distribution includes the Arduino IDE software in the standard software repository. You can install the Arduino IDE software using the standard Ubuntu Package Manager program. There are two packages to install: arduino and arduino-core. When you install these packages, Ubuntu automatically places a link to the Arduino IDE software in the Unity desktop menu.

Summary

In this hour, you learned how you can write code to program the Arduino. The core of the Arduino is the ATmega microcontroller, so the code you write must be able to run on that system. The ATmega microcontroller uses machine language code to handle data and interface with the input and output ports. However, you can use the Atmel C programming language library to make coding easier. Also, the Arduino development team has expanded on the Atmel C language and provided their own programming environment and libraries. This makes creating code for the Arduino much easier. You can download the Arduino IDE package from the Arduino website for the Windows, OS X, and Linux platforms.

In the next hour, we take a closer look at the Arduino IDE window. You'll learn how to create, edit, compile, debug, and upload your Arduino programs, all from one place!

Workshop

Quiz

1. Which part of the microcontroller performs the addition of two numbers?
    - **A.** The ALU
    - **B.** The registers
    - **C.** The memory
    - **D.** The I/O interface
2. SRAM memory stores your program data, even after you turn off the Arduino unit. True or false?
3. Why doesn't the Arduino have any analog output ports?

Answers

1. A. The Arithmetic Logic Unit (ALU) performs all mathematical operations on data in the microcontroller.
2. False. The SRAM memory loses its contents when the Arduino power is off. You must store any data you want to keep either in the EEPROM memory or in a removable storage device, such as an SD card.
3. The Arduino uses pulse-width modulation (PWM) to simulate an analog output signal using a digital output port. That allows you to use the digital output ports to generate either a digital or analog output.

Q&A

Q. Can I write assembly language programs and upload them to the Arduino?

A. Yes, you can write assembly language programs using the Arduino IDE and compile them using the avr-gcc compiler. However, you must be careful when doing that, because you have to keep track of the registers and memory locations in your program!

Q. Is the Atmel Studio IDE free?

A. Yes, the Atmel Studio IDE is free software, but it's not open source software.

HOUR 3

Using the Arduino IDE

What You'll Learn in This Hour:

- The Arduino IDE interface layout
- Using the features of the Arduino IDE menu bar
- Configuring the Arduino IDE to work with your Arduino
- Using the serial monitor feature in the Arduino IDE

The Arduino IDE software package makes it easier to create and work with Arduino programs. This hour covers the different parts of the Arduino IDE interface and how to use them to create your Arduino programs.

Overview of the IDE

The Arduino integrated development environment (IDE) itself is a Java program that creates an editor environment for you to write, compile, and upload your Arduino programs. Figure 3.1 shows the basic layout of the Arduino IDE interface.

The Arduino IDE interface contains five main sections:

- The menu bar
- The taskbar
- The editor window
- The message area
- The console window

Let's walk through each of these different sections and explain what they're for and how to use them.

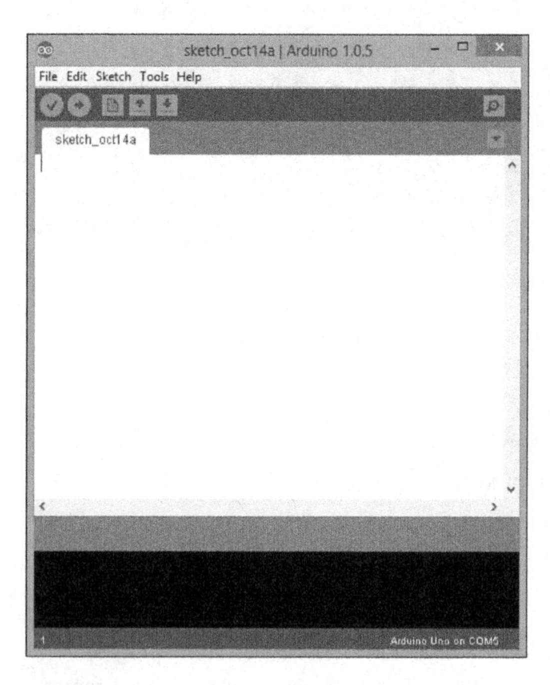

FIGURE 3.1
The Arduino IDE main window.

Walking Through the Menus

The menu bar provides access to all the features of the Arduino IDE. From here, you can create a new program, retrieve an existing program, automatically format your code, compile your program, and upload your program to the Arduino unit. The following sections discuss each of the options you'll find in the menu bar.

The File Menu

The File menu contains all the file-handling options required to load and save sketches, along with a couple of miscellaneous options that don't fit in any other menu bar category. The options available under the File menu are described in the following subsection.

New

The New option does what you'd probably expect it to do: It creates a new Arduino sketch tab in the main window. When you select the option to create a new sketch, the Arduino IDE automatically assigns it a name, starting with sketch_, followed by the month and date (such as jan01), followed by a letter making the sketch name unique. When you save the sketch, you can change this sketch name to something that makes more sense.

Open

The Open option opens an existing sketch using the File Open dialog box of the operating system. The Open dialog box will open in the default sketchbook folder for your Arduino IDE (set in the Preferences option, discussed later). The File Open dialog box allows you to navigate to another folder area if you've saved sketches on a removable media device, such as a USB thumb drive.

TIP

Working with Older Sketches

Older versions of the Arduino IDE saved sketch files using a different format and file extension name (`.pde`) than the 1.0 Arduino IDE series. The 1.0 series can open the older sketch files, but when you save the updated sketch, it will automatically convert it to the newer `.ino` format if you set that option in the Preferences.

Sketchbook

The Sketchbook option provides a quick way to open your existing sketches. It lists all the sketches that you've saved in the default sketchbook area in the Arduino IDE. You can select the sketch to load from this listing.

Examples

The Examples option provides links to lots of different types of example sketches provided by the Arduino developers. The examples are divided into different categories, so you can quickly find an example sketch for whatever type of project you're working on.

When you select an example sketch, the sketch code automatically loads into the editor window. You can then verify and upload the example sketch directly, or you can modify the sketch code before compiling and uploading.

CAUTION

Modifying Example Sketches

If you modify the example sketch and try to save it, the Arduino IDE will warn you that the example sketch is saved as a read-only file and not let you update it. If you want to save your modifications, use File > Save As from the menu bar to save the modified sketch as a different filename in your standard sketchbook folder.

Close

The Close option closes out the current sketch editor window and safely closes the sketch file. Closing the sketch will also exit the Arduino IDE interface.

Save

The Save option allows you to save a previously saved sketch in your sketchbook area using the same filename. Be careful when using this option, as the Arduino IDE will just overwrite the existing saved sketch and it won't keep any past versions automatically for you.

Save As

The Save As option allows you to save a sketch using a new filename. The Arduino IDE saves each sketch under a separate folder under the Arduino IDE sketchbook folder area. It saves the sketch file using the `.ino` file extension.

Upload

The Upload option uploads the compiled sketch code to the Arduino unit using the USB serial interface. Make sure that the Arduino unit is connected to the USB port before you try to upload the sketch.

Upload Using Programmer

The Upload Using Programmer option is for more advanced users. The normal Upload option uploads your sketch to a special location in the flash program memory area on the ATmega microcontroller. The Upload Using Programmer option writes the program to the start of the flash memory area on the ATmega microcontroller. This requires additional coding in your sketch so that the Arduino can boot from your program.

TIP

The Arduino Bootloader

The Arduino system adds a special bootloader program to the start of the microcontroller's flash memory area. The bootloader runs when the Arduino powers on, and automatically jumps to your program code. It makes formatting and running your Arduino code easier because you don't have to worry about the boot code.

However, the Arduino IDE is versatile enough to allow advanced users to upload programs directly to the program area on the microcontroller. When you do this, your program code is responsible for creating a fully operational program.

Beginners are best off using the Arduino bootloader format to write the program code. Not using the bootloader means that you need to include more code in your Arduino program to tell the microcontroller how to handle your program.

Page Setup

The Page Setup option allows you to define formatting options for printing your sketch on a printer connected to your workstation.

Print

The Print option allows you to send the sketch text to a printer connected to your workstation.

Preferences

The Preferences option displays a dialog box that contains several settings for the Arduino IDE that you can change:

- **Sketchbook Location:** Sets the folder used to store sketches saved from the Arduino IDE.
- **Editor Language:** Sets the default language recognized in the text editor.
- **Editor Font Size:** Sets the font size used in the text editor.
- **Show Verbose Output:** Displays more output during either compiling or uploading. This proves handy if you run into any problems during either of these processes.
- **Verify Code After Upload:** Verifies the machine code uploaded into the Arduino against the code stored on the workstation.
- **Use External Editor:** Allows you to use a separate editor program instead of the built-in editor in the Arduino IDE.
- **Check for Updates on Startup:** Connects to the main Arduino.cc website to check for newer versions of the Arduino IDE package each time you start the Arduino IDE.
- **Update Sketch Files to New Extension on Save:** If you have sketches created using older versions of the Arduino IDE, enable this setting to automatically save them in the new `.ino` format.
- **Automatically Associate .ino Files with Arduino:** Allows you to double-click `.ino` sketch files from the Windows Explorer program to open them using the Arduino IDE.

You can change these settings at any time, and they will take effect immediately in the current Arduino IDE window.

Quit

Closes the existing sketch and exits the Arduino IDE window.

The Edit Menu

The Edit menu contains options for working with the program code text in the editor window. The options available under the Edit menu are:

Undo

This feature enables you to take back changes that you make to the sketch code. This returns the code to the original version.

Redo

This feature enables you to revert back to the changes you made (undoes an Undo command).

Cut

This feature enables you to cut selected text from the sketch code to the system Clipboard. The selected text disappears from the editor window.

Copy

This feature allows you to copy selected text from the sketch code to the system Clipboard. With the Copy feature, the selected text remains in place in the sketch code.

Copy for Forum

This interesting option is for use if you interact in the Arduino community forums. If you post a problem in the forums, members will often ask to see your code. When you use this feature, the Arduino IDE copies your sketch code to the Clipboard and retains the color formatting that the editor uses to highlight Arduino functions. When you paste the code into your forum message, it retains that formatting, making it easier for other forum members to follow your code.

Copy as HTML

The Copy as HTML option also keeps the color and formatting features of the editor, but embeds them as HTML code in your program code. This option comes in handy if you want to post your sketch code onto a web page to share with others.

CAUTION

HTML Code in Sketches

Be careful when using this feature, because it embeds HTML tags into your sketch code. You can't run the code this option generates in your Arduino IDE editor; it's meant only for pasting into a web page HTML document.

Paste

As you might expect, the Paste option copies any text currently in your system Clipboard into the editor window.

Select All

Yet another common editor feature, the Select All option highlights all the text in the editor window to use with the Copy or Cut options.

Comment/Uncomment

You use this handy feature when troubleshooting your sketch code. You can highlight a block of code and then select this option to automatically mark the code as commented text. The compiler ignores any commented text in the sketch code.

When you finish troubleshooting, you can highlight the commented code block and select this option to remove the comment tags and make the code active again.

Increase Indent

This option helps format your code to make it more readable. It increases the space used to indent code blocks, such as `if-then` statements, `while` loops, and `for` loops. This helps the readability of your code.

Decrease Indent

This option enables you to reduce the amount of space placed in front of indented code blocks. Sometimes if code lines run too long, it helps to reduce the indentation spaces so that the code more easily fits within the editor window.

Find

The Find option provides simple search and replace features in the editor window. If you just want to find a specific text string in the code, enter it into the Find text box. You can select or deselect the Ignore Case check box as required. The Wrap Around check box helps you find longer text strings that may have wrapped around to the next line in the editor window.

If you need to quickly replace text, such as if you're changing a variable name in your sketch code, you can enter the replacement text in the Replace With text box. You can then replace a single instance of the found text, or click the Replace All button to replace all occurrences of the text with the replacement text.

Find Next

The Find Next option finds the next occurrence of the previously submitted search text.

Find Previous

The Find Previous option finds the previous occurrence of the previously submitted search text.

Use Selection for Find

This option lets you control just what section of your code the Find feature looks in. Instead of searching an entire code file, you can highlight a section of the code and use this feature to find a text string just within the highlighted section.

The Sketch Menu

The Sketch menu provides options for working with the actual sketch code. The following subsections describe the various options available under the Sketch menu.

Verify/Compile

The Verify/Compile option checks the Arduino program for any syntax errors, and then processes the code through the avr-gcc compiler to generate the ATmega microcontroller machine code. The result of the Verify/Compile process displays in the console window of the Arduino IDE. If any errors in the program code generate compiler errors, the errors will appear in the console window.

Show Sketch Folder

The Show Sketch Folder option shows all the files stored in the folder associated with the sketch. This comes in handy if you have multiple library files stored in the sketch folder.

Add File

The Add File menu option enables you to add a library file to the sketch folder.

Import Library

The Import Library menu option automatically adds a C language `#include` directive related to the specific library that you select. You can then use functions from that library in your program code.

The Tools Menu

The Tools menu provides some miscellaneous options for your sketch environment, as described in the following subsections.

Auto Format

The Auto Format option helps tidy up your program code by indenting lines of code contained within a common code block, such as `if-then` statements, `while` loops, and `for` loops. Using the Auto Format feature helps make your program code more readable, both for you and for anyone else reading your code.

Archive Sketch

This feature saves the sketch file as an archive file for the operating system platform used. For Windows, this appears as a compressed folder using the zip format. For Linux, this appears as a `.tar.gz` file.

Fix Encoding and Reload

This option is one of the more confusing options available in the Arduino IDE. If you load a sketch file that contains non-ASCII characters, you'll get odd-looking characters appear in the editor window where the non-ASCII characters are located. When you select this option, the Arduino IDE reloads the file and saves it using UTF-8 encoding.

Serial Monitor

The Serial Monitor option enables you to interact with the serial interface on the Arduino unit. It produces a separate window dialog box, which displays any output text from the Arduino and allows you to enter text to send to the serial interface on the Arduino. You must write the code in your programs to send and receive the data using the Arduino USB serial port for this to work.

Board

The Board option is extremely important. You must set the Board option to the Arduino unit that you're using so that the Arduino IDE compiler can create the proper machine code from your sketch code.

Serial Port

The Serial Port option selects the workstation serial port to use to upload the compiled sketch to the Arduino unit. You must select the serial port that the Arduino is plugged into.

Programmer

The Programmer option is for advanced users to select what version of code to upload to the ATmega microcontroller. If you use the Arduino bootloader format (which is recommended), you don't need to set the Programmer option.

Burn Bootloader

This option enables you to upload the Arduino bootloader program onto a new ATmega microcontroller. If you purchase an Arduino unit, this code is already loaded onto the microcontroller installed in your Arduino, so you don't need to use this feature.

However, if you have to replace the ATmega microcontroller chip in your Arduino, or if you purchase a new ATmega microcontroller to use in a separate breadboard project, you must burn the bootloader program into the new microcontroller unit before you can use the Arduino IDE program format for your sketches.

CAUTION

Working with the Bootloader

All the current Arduino units load the bootloader program into the ATmega microcontroller program memory by default. If you purchase an Arduino unit, you won't have to worry about using the Burn Bootloader feature. However, if you purchase a new ATmega microcontroller to use outside of your Arduino, you may or may not have to burn the bootloader program. Some vendors preload the bootloader program for you, others don't. Check with your vendor to find out which option they use. If the bootloader isn't burned onto the microcontroller, your Arduino programs won't work correctly.

The Help Menu

The Help menu provides links to help topics for the Arduino IDE. The Help menu consists of six sections that relate to the Arduino IDE documentation:

- Getting Started
- Environment
- Troubleshooting
- Reference
- Find in Reference
- Frequently Asked Questions

Besides these six topics, two links provide information about the Arduino IDE:

- Visit Arduino.cc
- About Arduino

Exploring the Toolbar

If you don't want to have to navigate your way around the menu bar to perform functions in the Arduino, you can use a handy toolbar just under the menu bar. The toolbar provides icons for some of the more popular menu bar functions:

- **Verify (the checkmark icon):** Performs the Verify/Compile option.
- **Upload (the right arrow icon):** Uploads the compiled sketch code to the Arduino unit.
- **New (the page icon):** Opens a new sketch tab in the Arduino IDE window.
- **Open (the up arrow icon):** Displays the File Open dialog box that allows you to select an existing sketch to load into the editor.
- **Save (the down arrow icon):** Saves the sketch code into the current sketch filename.
- **Serial monitor (the magnifying glass icon):** Opens the serial monitor window to interact with the serial port on the Arduino unit.

Now that you've seen the menu bar and toolbar options, the next section walks through the basics of setting up the Arduino IDE to work with your Arduino unit.

Exploring the Message Area and Console Window

The last two sections of the Arduino IDE are provided to give you some idea of what's going on when you compile or upload your sketch. The message area is a one-line section of the Arduino IDE interface that displays quick messages about what's happening. For example, as you compile the sketch, it shows a progress bar indicating the progress of the compile process, as shown in Figure 3.2.

The console window often displays informational messages from the last command you entered. For example, after a successful compile, you'll see a message showing how large the program is, as shown in Figure 3.3.

If there are any errors in the compiled sketch code, both the message area and the console window will display messages pointing to the line or lines that contain the errors.

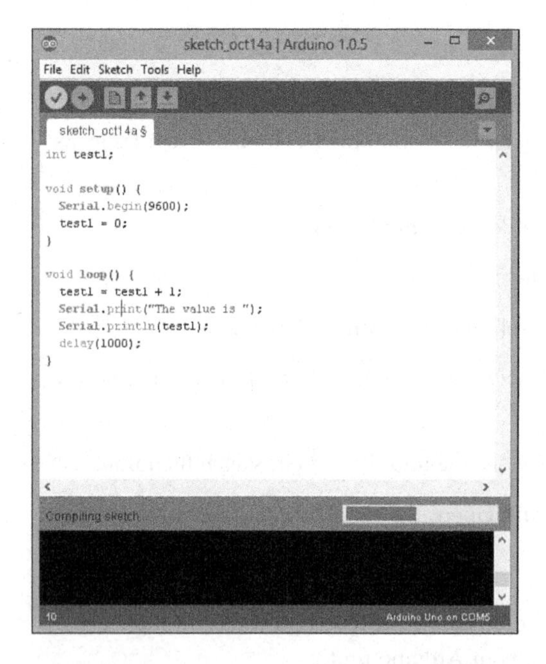

FIGURE 3.2
The Arduino IDE message area indicating the compile progress.

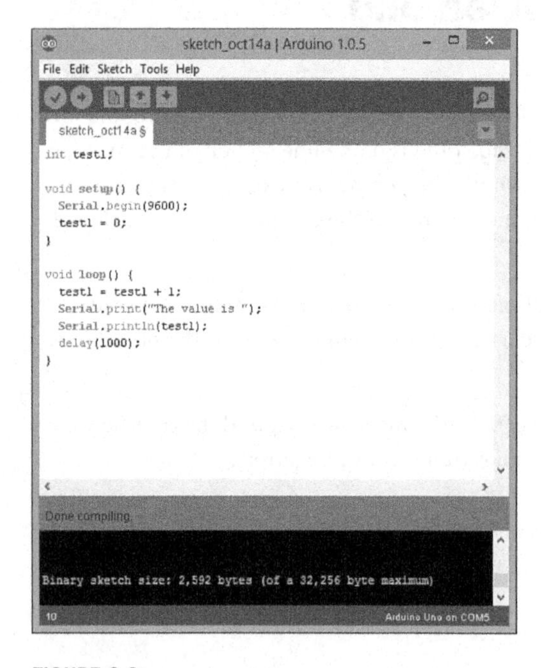

FIGURE 3.3
The Arduino IDE message window displaying the compiler result.

Setting Up the Arduino IDE

You should be able to work with your Arduino IDE with most of the default settings. However, you'll need to check on two main things before you start working with your Arduino unit. Follow these steps to get your Arduino IDE ready to work with your specific Arduino unit.

TRY IT YOURSELF ▼

Setting Up the Arduino IDE

1. Make sure that the Arduino IDE is set for your specific Arduino unit. Choose Tools > Board from the menu bar. This provides a list of the various Arduino boards available, as shown in Figure 3.4.

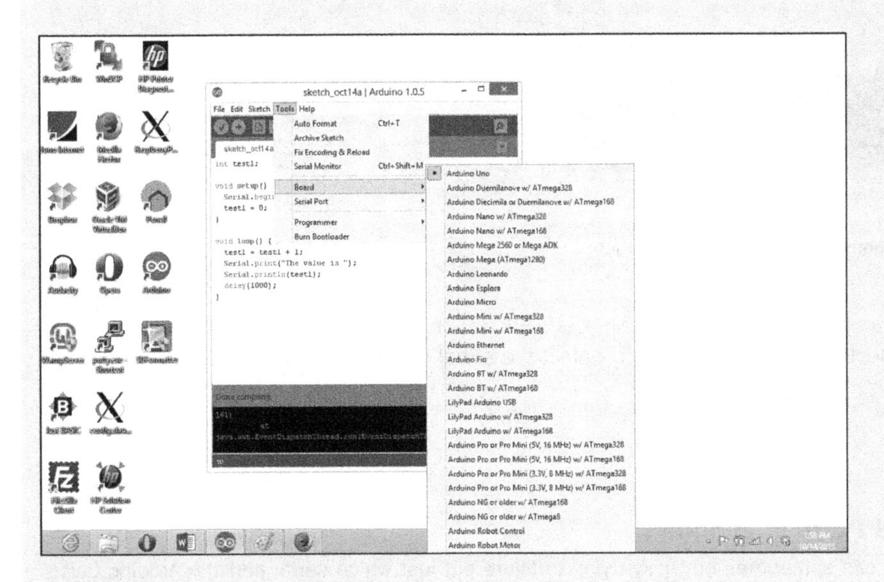

FIGURE 3.4
The Arduino board options in the Arduino IDE.

2. Select the board model that you're using to ensure the compiler generates the proper machine code for your Arduino unit.

3. The second thing to check is the serial port. Choose Tools > Serial Port from the menu bar. You should see a list of the available serial ports installed on your workstation, as shown in Figure 3.5.

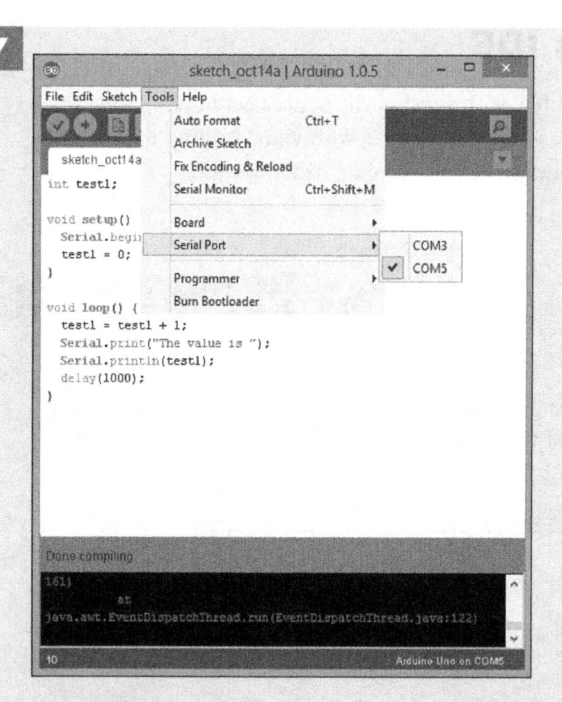

FIGURE 3.5
The serial port options for a Windows workstation.

4. The serial ports listed will depend on your workstation configuration, and what devices you have connected to your workstation. Select the serial port that points to your Arduino IDE.

TIP

Finding Serial Ports in Windows

With Windows, it can sometimes be tricky trying to figure out just which serial port the Arduino unit uses. The easiest way to determine the serial port is to open the Device Manager utility in Windows. From the Device Manager utility, you should see a Ports section. Click the plus sign (+) to expand that section, and it should list the COM port the Arduino unit is assigned to.

Using the Serial Monitor

The serial monitor is a special feature in the Arduino IDE that can come in handy when troubleshooting code running on your Arduino, or just for some simple communication with the program running on the Arduino.

The serial monitor acts like a serial terminal, enabling you to send data to the Arduino serial port and receive data back from the Arduino serial port. The serial monitor displays the data it receives on the selected serial port in a pop-up window.

You can activate the serial monitor feature in the Arduino IDE in three ways:

- Choose Tools > Serial Monitor from the menu bar.
- Press the Ctrl+Shift+M key combination.
- Click the serial monitor icon on the toolbar.

If you've selected the correct serial port for your Arduino unit, a pop-up window will appear and connect to the serial port to communicate with the Arduino unit serial port using the USB connection. Figure 3.6 shows the serial monitor window.

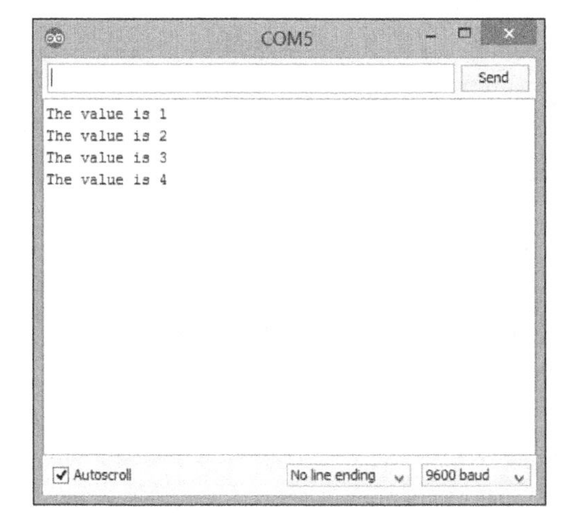

FIGURE 3.6
The serial monitor interface window.

You can enter text in the top line of the serial monitor interface and then click the Send button to send it to the Arduino unit via the serial port. The serial monitor window also displays any text output from the Arduino unit in a scrollable window area.

Three setting options are available at the bottom of the serial monitor window:

- **Autoscroll:** Checking this option always displays the last line in the scroll window. Removing the check freezes the window so that you can manually scroll through the output.

- **Newline:** This option controls what type of end-of-line marking the serial monitor sends after the text that you enter. You can choose to send no end-of-line marking, a standard UNIX-style newline character, a carriage control character, or a Windows-style newline and carriage control characters. Which you select depends on how you write your Arduino sketch to look for data.
- **Baud Rate:** The communications speed that the serial monitor connects to the Arduino serial port. By default, the serial monitor will detect the baud rate based on the `Serial.begin()` function used in the sketch code.

CAUTION

Starting the Serial Monitor

When you start the serial monitor, it automatically sends a reboot signal to the Arduino unit, causing it to restart the program that's currently loaded into the flash memory.

Summary

This hour covered the Arduino IDE user interface. You learned about each of the menu bar and taskbar options, including how to create, save, compile, and upload your Arduino sketches. After that, we looked at the serial monitor feature in the Arduino IDE. The serial monitor enables you to send and receive text with the Arduino unit using the USB serial port. This is a handy feature when you're trying to debug Arduino program code.

In the next hour, we create an Arduino program so that you can get a feel for the Arduino IDE and for working with your Arduino unit.

Workshop

Quiz

1. Which tool should you use to send data to the Arduino unit?
   - **A.** Message area
   - **B.** Console window
   - **C.** Serial monitor
   - **D.** Toolbar

2. The upload icon (the right arrow icon) on the toolbar will automatically verify and compile the sketch code before uploading it. True or false?

3. How does the Arduino IDE know what USB port to use to communicate with the Arduino device?

Answers

1. C. The serial monitor enables you to send and receive data from the Arduino unit.

2. True. Clicking the upload toolbar icon will first verify and compile the sketch code before trying to upload it, even if you've already compiled the code previously.

3. You must select the serial port (after choosing Tools > Serial Port from the menu bar) before you try to upload your sketch code. The Arduino unit will appear as an available serial port in the listing.

Q&A

Q. Is the Arduino IDE interface different if I load it in Windows, OS X, or Linux?

A. No, the Arduino IDE is written using the Java programming language, and it creates a common window interface on all three operating system platforms.

Q. Can I upload the sketch code to my Arduino unit using a wireless connection?

A. No, at this time you can only upload the sketch code using a serial USB port connection.

Q. Will the Arduino IDE editor window point out syntax errors in my sketch code before I try to compile it?

A. Sometimes. The Arduino IDE editor will highlight library function names and text that it thinks is part of a string, but it can't pick up all syntax errors in your code.

HOUR 4
Creating an Arduino Program

What You'll Learn in This Hour:

- Building an Arduino sketch
- Compiling and running a sketch
- Interfacing your Arduino to electronic circuits

Now that you've seen what the Arduino is and how to program it using the Arduino IDE, it's time to write your first program and watch it work. In this hour, you learn how to use the Arduino IDE software package to create, compile, and upload an Arduino program. You then learn how to interface your Arduino with external electronic circuits to complete your Arduino projects.

Building an Arduino Sketch

Once you have your Arduino development environment set up, you're ready to start working on projects. This section covers the basics that you need to know to start writing your sketches and getting them to run on your Arduino.

Examining the Arduino Program Components

When you use the Arduino IDE package, your sketches must follow a specific coding format. This coding format differs a bit from what you see in a standard C language program.

In a standard C language program, there's always a function named `main` that defines the code that starts the program. When the CPU starts to run the program, it begins with the code in the `main` function.

In contrast, Arduino sketches don't have a `main` function in the code. The Arduino bootloader program that's preloaded onto the Arduino functions as the sketch's `main` function. The Arduino starts the bootloader, and the bootloader program starts to run the code in your sketch.

The bootloader program specifically looks for two separate functions in the sketch:

- `setup`
- `loop`

The Arduino bootloader calls the `setup` function as the first thing when the Arduino unit powers up. The code you place in the `setup` function in your sketch only runs one time; then the bootloader moves on to the `loop` function code.

The `setup` function definition uses the standard C language format for defining functions:

```
void setup() {
code lines
}
```

Just place the code you need to run at startup time inside the `setup` function code block.

After the bootloader calls the `setup` function, it calls the `loop` function repeatedly, until you power down the Arduino unit. The `loop` function uses the same format as the `setup` function:

```
void loop() {
code lines
}
```

The meat of your application code will be in the `loop` function section. This is where you place code to read sensors and send output signals to the outputs based on events detected by the sensors. The `setup` function is a great place to initialize input and output pins so that they're ready when the loop runs, then the `loop` function is where you use them.

Including Libraries

Depending on how advanced your Arduino program is, you may or may not need to use other functions found in external library files. If you do need to use external libraries, you first need to define them at the start of your Arduino program, using the `#include` directive:

```
#include <library>
```

The `#include` directives will be the first lines in your sketch, before any other code.

If you're using a standard Arduino shield, most likely the shield library code is already included in the Arduino IDE package. Just choose Sketch > Import Library from the menu bar, and then select the shield that you're using. The Arduino IDE automatically adds the `#include` directives required to write code for the requested shield. For example, if you select the Ethernet shield, the following lines are imported into the sketch:

```
#include <Dhcp.h>
#include <Dns.h>
#include <Ethernet.h>
#include <EthernetClient.h>
#include <EthernetServer.h>
#include <EthernetUdp.h>
#include <util.h>
```

That saves a lot of time from having to go hunting around to find the libraries required for a specific shield.

Creating Your First Sketch

Now that you've seen the basics for creating an Arduino program, let's dive in and create a simple sketch to get a feel for how things work.

Working with the Editor

When you open the Arduino IDE, the editor window starts a new sketch. The name of the new sketch appears in the tab at the top of the editor window area, in the following format:

`sketch_`*`mmmddx`*

where *mmm* is a three-letter abbreviation of the month, *dd* is the two-digit numerical date, and *x* is a letter to make the sketch name unique for the day (for example, sketch_jan01a).

As you type your sketch code into the editor window, the editor will color-code different parts of the sketch code, such as making function names brown and text strings blue. This makes it easier to pick out syntax errors, and comes in handy when you're trying to debug your sketch.

Now you're ready to start coding. Listing 4.1 shows the code for the sketch0401 file that we'll use to test things out. Enter this code into the Arduino IDE editor window.

LISTING 4.1 The sketch0401 Code

```
int counter = 0;
int pin = 13;

void setup() {
  Serial.begin(9600);
  pinMode(pin, OUTPUT);
  digitalWrite(pin, LOW);
}

void loop() {
  counter = counter + 1;
```

```
  digitalWrite(pin, HIGH);
  Serial.print("Blink #");  Serial.println(counter);
  delay(1000);
  digitalWrite(pin, LOW);
  delay(1000);
}
```

You'll learn what all these different lines of code mean as you go through the rest of the hours, so don't worry too much about the code for now. The main point now is to have a sketch to practice compiling and running.

The basic idea for this code is to make the Arduino blink the L LED connected to digital port 13 on the Arduino once per second, and also output a message to the Arduino serial port, counting each blink.

After you enter the code into the editor window, choose File > Save As from the menu bar to save the sketch as `sketch0401`. Now you're ready to verify and compile the sketch.

Compiling the Sketch

The next step in the process is to compile the sketch code into the machine language code that the Arduino runs.

Click the verify icon on the toolbar (the checkmark icon), or choose Sketch > Verify/Compile from the menu bar. Figure 4.1 shows the results that you should get if things worked correctly.

As shown in Figure 4.1, you should see a message in the message area that the compile has completed, and the console window should show the final size of the compiled machine language code that will be uploaded to the Arduino.

If you have any typos in the sketch code that cause the compile process to fail, you'll see an error message in the message area, as shown in Figure 4.2.

The Arduino IDE also highlights the line of code that generated the error, making it easier for you to pick out the problem. Also, a more detailed error message appears in the console window area to help even more.

After you get the sketch to compile without any errors, the next step is to upload it to your Arduino.

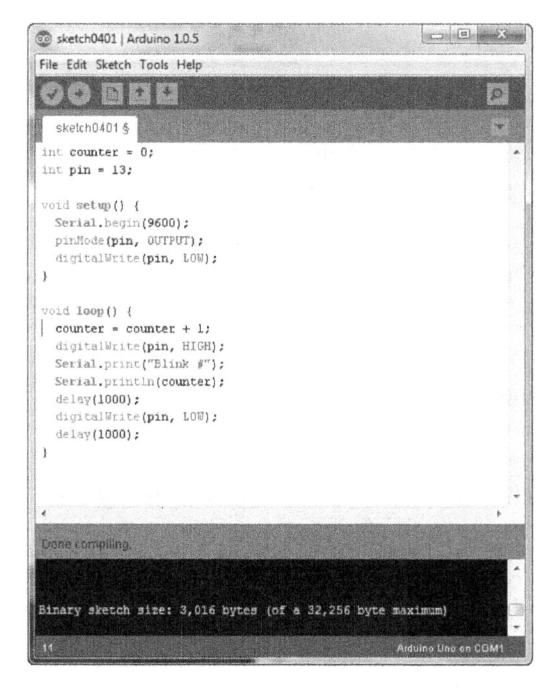

FIGURE 4.1
Compiling the sketch0401 code.

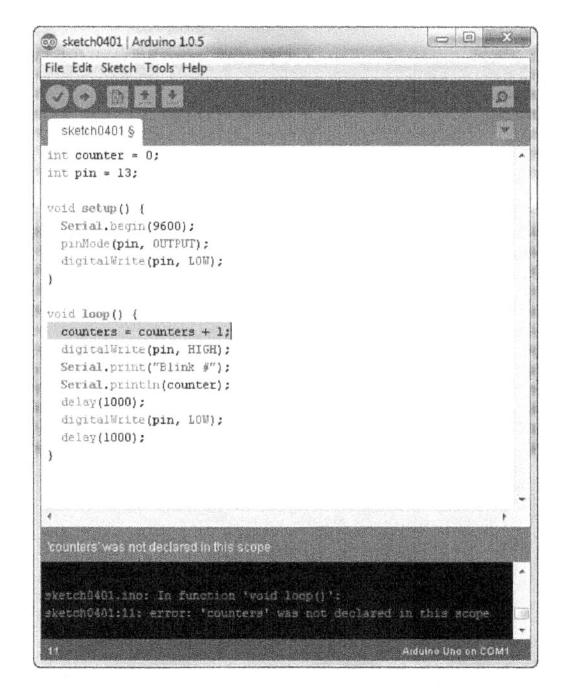

FIGURE 4.2
A compiler error displayed in the Arduino IDE.

Uploading Your Sketch

The key to successfully uploading sketches to your Arduino unit is in defining how the Arduino is connected to your workstation. Hour 3, "Using the Arduino IDE," walked through how to use the Tools > Serial Port menu bar option to set which serial port your Arduino is connected to. After you set that, you should be able to easily upload your compiled sketches.

Just click either the upload icon on the toolbar (the right arrow icon), or select File > Upload from the menu bar. Before the upload starts, the Arduino IDE recompiles the sketch code. This comes in handy when you're just making quick changes; you can compile and upload the new code with just one click.

When the upload starts, you should see the TX and RX LEDs on the Arduino blink, indicating that the data transfer is in progress. When the upload completes, you should see a message in both the Arduino IDE message area and console window indicating that the upload was completed. If anything does go wrong, you'll see an error message appear in both the message area and the console window, as shown in Figure 4.3.

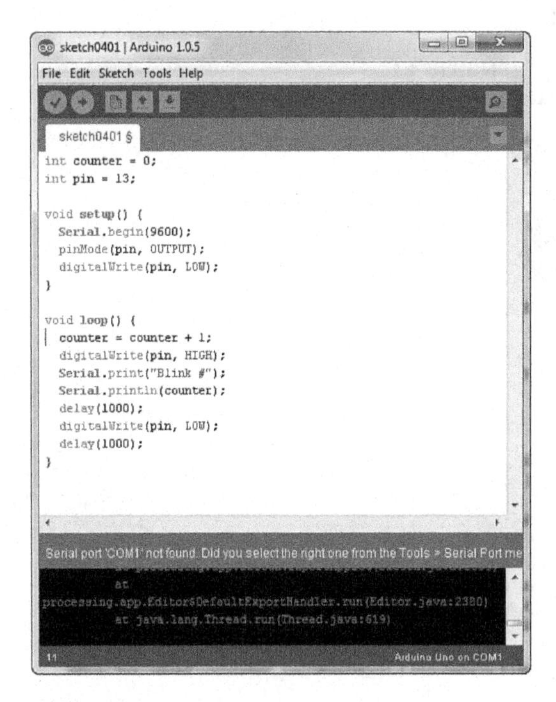

FIGURE 4.3
Upload problem message in the Arduino IDE.

If all goes well, you're ready to start running your sketch on the Arduino. The next section shows you how.

Running Your Program

Now that the sketch code is uploaded onto your Arduino, you're ready to start running it. However, you might have noticed that once the upload process finished in the Arduino IDE, the L and the TX LEDs on your Arduino unit already started to blink. That's your sketch running. When the upload process completes, the bootloader automatically reboots the Arduino and runs your program code.

The L LED is blinking because of the `digitalWrite()` function setting the digital pin 13 first to 0 (no voltage) and then after a second, setting it to 1 (producing a 5V signal). The TX LED is blinking because the `Serial.print()` function is sending data out the serial port.

You can view the output from the serial port on your Arduino using the serial monitor built in to the Arduino IDE. Just choose Tools > Serial Monitor from the menu bar, or click the serial monitor icon (the magnifying glass icon) on the toolbar. The serial monitor window appears and displays the output received from the Arduino, as shown in Figure 4.4.

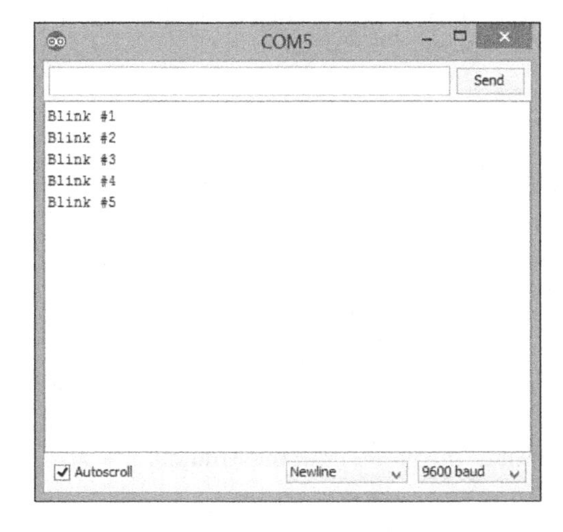

FIGURE 4.4
Viewing the Arduino serial port output from the serial monitor.

You might have noticed that after you started the serial monitor, the blink count output restarted back at 1. When you start serial monitor, it sends a signal to the Arduino to reset it, which in turn runs the bootloader to reload the sketch and start over from the beginning.

You can also manually restart a running sketch using the Reset button on the Arduino. On the Arduino Uno R3, you'll find the Reset button in the upper-left corner of the circuit board. Just push the button and release it to reset the Arduino.

You don't have to connect the Arduino to the USB port on your workstation for it to run. You can run the Arduino from an external power source, as well, such as a battery pack or AC/DC converter. Just plug the power source into the power socket on the Arduino unit. The Arduino Uno R3 automatically detects power applied to either the USB port or the power port and starts the bootloader program to start your sketch.

Interfacing with Electronic Circuits

While getting your sketch uploaded to the Arduino and running is a significant accomplishment, most likely you'll want to do more in your Arduino projects than just watch the L LED blink. That's where you'll need to incorporate some type of external electronic circuits into your projects. This section covers the basics of what you need to know to add external electronic circuits to your Arduino sketches.

Using the Header Sockets

The main use of the Arduino is to control external electronic circuits using the input and output signals. To do that, you need to interface your electronic circuits with the Arduino analog and digital signals. This is where the header sockets come into play.

If you remember from Hour 1, "Introduction to the Arduino," the header sockets are the two rows of sockets at the top and bottom of the Arduino Uno circuit board. (Some more advanced Arduino units, such as the Arduino Mega, also include a third header socket on the right side of the board to support additional ports.) You'll plug your electronic circuits into the sockets to gain access to the Arduino input and output signals, as well as the power from the Arduino.

The basic Arduino Uno unit that we're using for our experiments uses the standard Arduino two-row header socket format. Figure 4.5 shows the layout of the upper and lower header sockets.

FIGURE 4.5
The Arduino Uno upper and lower header sockets.

The lower header socket has 13 ports on it, as described in Table 4.1.

TABLE 4.1 The Arduino Uno Lower Header Socket Ports

Label	Description
IOREF	Provides the reference voltage used by the microcontroller if not 5V.
RESET	Resets the Arduino when set to LOW.
3.3V	Provides a reduced 3.3V for powering low-voltage external circuits.
5V	Provides the standard 5V for powering external circuits.
GND	Provides the ground connection for external circuits.
GND	A second ground connection for external circuits.
Vin	An external circuit can supply 5V to this pin to power the Arduino, instead of using the USB or power jacks.
A0	The first analog input interface.
A1	The second analog input interface.

Label	Description
A2	The third analog input interface.
A3	The fourth analog input interface.
A4	The fifth analog input interface, also used as the SDA pin for TWI communications.
A5	The sixth analog input interface, also used as the SCL pin for TWI communications.

The upper header socket has 16 ports on it, as described in Table 4.2

TABLE 4.2 The Arduino Uno Upper Header Socket Ports

Label	Description
AREF	Alternative reference voltage used by the analog inputs (by default, 5V).
GND	The Arduino ground signal.
13	Digital port 13, and the SCK pin for SPI communication.
12	Digital port 12, and the MISO pin for SPI communication.
~11	Digital port 11, a PWM output port, and the MOSI pin for SPI communications.
~10	Digital port 10, a PWM output port, and the SS pin for SPI communication.
~9	Digital port 9, and a PWM output port.
8	Digital port 8.
7	Digital port 7.
~6	Digital port 6, and a PWM output port.
~5	Digital port 5, and a PWM output port.
4	Digital port 4.
~3	Digital port 3, and a PWM output port.
2	Digital port 2.
TX -> 1	Digital port 1, and the serial interface transmit port.
RX <- 0	Digital port 0, and the serial interface receive port.

For our test sketch, we need to access the digital port 13 socket, in addition to a GND socket, to complete the electrical connection to power our electronic devices.

To access the sockets, you can plug wires directly into the socket ports. To make it easier, you can use jumper wires, which you can easily remove when you finish experimenting.

Building with Breadboards

When you build an electronic circuit, the layout is usually based on a schematic diagram that shows how the components should be connected. The schematic shows a visual representation of which components are connected to which, using standard symbols to represent the different components, such as resistors, capacitors, transistors, switches, relays, sensors, and motors.

Your job is to build the electronic circuit to mimic the layout and connections shown in the schematic diagram. In a permanent electronic circuit, you use a printed circuit board (called PCB) to connect the components according to the schematic.

In a PCB, connections between the electronic components are etched into the PCB using a metallic conductor. To place the electronic components onto the PCB, you must solder the leads of the components onto the PCB.

The downside to using a PCB for your electronic project is that because it's intended to be permanent, you can't easily make changes. Although that's fine for final circuits, when you're developing a new system and experimenting with different circuit layouts, it's somewhat impractical to build a new PCB layout for each test.

This is where breadboards come in handy. A breadboard provides an electronic playground for you to connect and reconnect electronic components as you need. Figure 4.6 shows a basic breadboard layout.

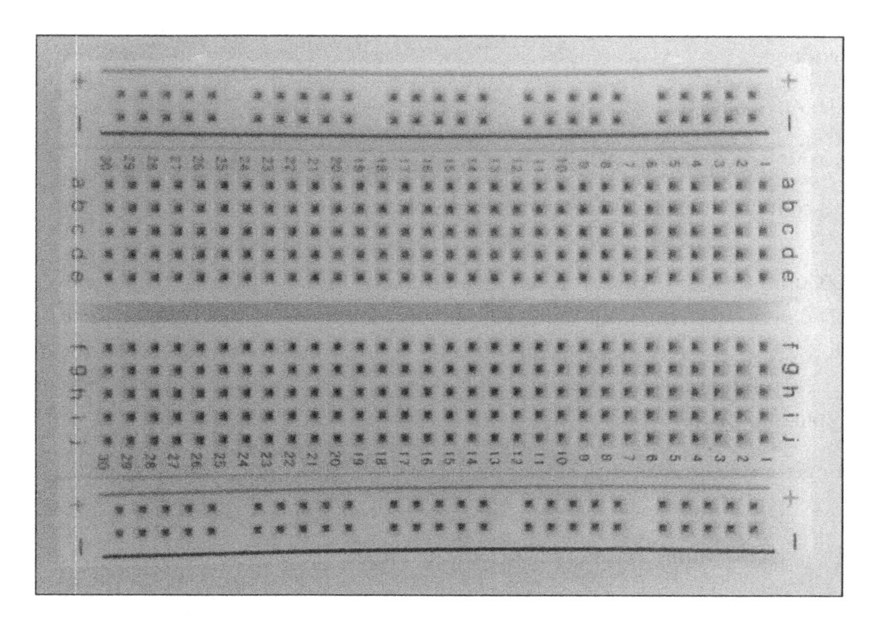

FIGURE 4.6
A basic breadboard.

Breadboards come in many different sizes and layouts, but most breadboards have these features:

- A long series of sockets interconnected along the ends of the breadboard. These are called buses (or sometimes rails), and are often used for the power and ground voltages. The sockets in the bus are all interconnected to provide easy access to power in the circuit.
- A short series of sockets (often around five) interconnected, and positioned across a gap in the center of the breadboard. Each group of sockets is interconnected to provide an electrical connection to the components plugged into the same socket group. The gap allows you to plug integrated circuit chips into the breadboard and have access to the chip leads.

The breadboard allows you to connect and reconnect your circuits as many times as you need to experiment with your projects. Once you get your circuit working the way you want, you can transfer the breadboard layout onto a PCB for a more permanent solution.

Adding a Circuit to Your Project

Now that you've seen how to add external electronic circuits to your Arduino project, let's create a simple circuit to add to our Arduino sketch. Instead of using the L LED on your Arduino, let's use an external LED.

For this project, you need the following parts:

- A standard breadboard (any size)
- A standard LED (any color)
- A 1000ohm resistor (color code brown, black, red)
- Jumper wires to connect the breadboard circuit to the Arduino

The circuit uses a 1000ohm resistor to limit the voltage that flows through the LED to help protect the LED. The LED doesn't need the full 5V provided by the Arduino output, so by placing a resistor in series with the LED, the resistor helps absorb some of the voltage, leaving less for the LED. If you don't have a 1000ohm resistor handy, you can use any other resistor value to help lessen the voltage applied to the LED.

Figure 4.7 shows connecting the resistor and LED to the GND and digital pin 13 ports on your Arduino Uno unit.

Just follow these steps to create your electronic circuit for the project.

FIGURE 4.7
Circuit diagram for the sample project.

TRY IT YOURSELF 

Creating the Electronic Circuit

1. Connect a jumper wire from one of the GND socket ports on the Arduino to a socket row on the breadboard.
2. Connect a jumper wire from the digital pin 13 socket port on the Arduino to another socket row on the breadboard (not the same as the one you used for the GND signal).
3. Plug the LED into the breadboard so that the longer lead of the LED is connected to the same socket row as the digital pin 13 wire and so that the other lead is plugged into a separate socket row on the breadboard.

CAUTION

Polarity in Electronic Circuits

While plugging the LED in the wrong way won't harm the LED, there are other electronic components that can cause damage if plugged in the wrong way (such as transistors). Be careful when working with electronic components that have polarity requirements!

4. Plug the resistor so that one lead connects to the same socket row as the short lead of the LED and so that the other lead connects to the socket row that carries the Arduino GND signal.

Now you should be ready to test things out. Power up the Arduino, either by connecting it to the USB port of your workstation or by connecting it to an external power source. Because the Arduino maintains the sketch in flash memory, you don't need to reload your sketch; it should start running automatically.

CAUTION

Providing Power to the Arduino

Be careful when plugging and unplugging your Arduino if you're using a USB hub with other devices. Stray voltages can result that may damage the other USB devices on the hub. It's always a good idea to power down your USB hub when plugging and unplugging the Arduino.

If things are working, you should see the LED on the breadboard blink once per second. If not, double-check your wiring to ensure that you have everything plugged together correctly on the breadboard and that the wires are plugged into the proper socket ports on the Arduino.

TIP

Using the Serial Monitor

If you connected the Arduino to the USB port on your workstation, you can still use the serial monitor in the Arduino IDE to view the output from the sketch. However, if you use an external power source to power the Arduino, you won't be able to view that output unless you connect an external serial device to the Arduino serial ports, which are digital ports 0 and 1 in the header sockets.

Summary

This hour walked you through your first Arduino project. First, we entered the sketch code into the Arduino IDE editor window, then we compiled the sketch, and finally, we uploaded the compiled sketch to the Arduino. You also saw how to use the serial monitor feature in the Arduino IDE to monitor output from your sketch. After that, you learned how to set up an external electronic circuit and interface it with your Arduino.

In the next hour, we take a closer look at the actual Arduino sketch code that we'll be using in our projects. You'll learn how the Arduino programming language stores and manipulates data within our sketches.

Workshop

Quiz

1. Which function must your Arduino sketch define to run the main part of your program code?
    - **A.** `setup`
    - **B.** `loop`
    - **C.** `main`
    - **D.** `start`
2. The Arduino IDE editor uses the same text color code to indicate functions as it does regular text in the code. True or false?
3. How do you interface external electronic circuits to your Arduino?

Answers

1. B. The `loop` function contains the sketch code that continually runs while the Arduino unit is powered on. This is where you need to place your main sketch code.
2. False. The Arduino IDE uses brown to indicate functions used in the sketch code, and uses blue to indicate text strings contained in the sketch code.
3. The Arduino header sockets are designed to easily interface external electronic circuits with the analog and digital input and output pins on the microcontroller.

Q&A

Q. Is there a limit to the size of the sketches I can upload to my Arduino?

A. Yes, the size of the sketch is limited by the amount of flash memory present on your Arduino. The Arduino Uno R3 has 32KB of flash memory. When you compile your sketch, the Arduino IDE console window will display the size of the compiled sketch code and how much space is remaining in the flash memory.

Q. Can I damage my Arduino by plugging in the wrong wires to the wrong header socket ports?

A. Yes, it's possible, but the Arduino does contain some basic protections. The Arduino is designed with some basic voltage protection on each of the input and output ports. If you supply too large of voltages to the ports, however, you can risk burning out the microcontroller chip. Use caution when connecting wires to the Arduino header sockets, and *always* double-check your work before turning on the power.

Q. Is there an easy way to identify resistor values when working with electronic circuits?

A. Yes, all resistor manufacturers use a standard resistor color code. The resistor value and tolerance are indicated by color bands around the resistor. To find the value of a resistor, refer to a resistor color-code chart, as shown in the Wikipedia article on electronic color codes (http://en.wikipedia.org/wiki/Electronic_color_code).

PART II

The C Programming Language

HOUR 5
Learning the Basics of C

What You'll Learn in This Hour:

- How to store your sketch data in variables
- How to use variables in mathematical operations
- How to output variable values to the Arduino serial port
- How to use some of the built-in Arduino C functions

This hour dives head first into the C programming language basics, which is the foundation of the Arduino programming language. It first explores how to store data in your sketches using variables. Then it discusses how to perform simple arithmetic operations in your sketches. The hour wraps up by going through some of the more common functions available in the standard Arduino library for you to use.

Working with Variables

If you remember from Hour 2, "Creating an Arduino Programming Environment," the Arduino programming language provides a user-friendly interface to write programs for the underlying ATmega microcontroller on the Arduino. One of those interfaces is how data is stored in memory.

Instead of having to reference specific memory locations to store your program data (like you do in assembly language), the Arduino programming language uses simple variables to represent memory locations for storing data. This section discusses how to use those variables in your Arduino sketches to store and retrieve data.

Declaring Variables

Because the Arduino programming language is built on the C programming language, it uses the standard C language format for creating and using variables. In the C programming language, creating a variable requires two steps:

1. Declare the variable for use in the sketch.
2. Assign a data value to the variable.

When you declare a variable, the C language compiler stores the variable name, along with some other information, in an internal table. This table allows the compiler to know what variables the program uses and how much space in memory they'll require when you run the program. However, when you declare a variable, the compiler doesn't actually assign the variable to a specific location in the system memory yet; that part comes later.

To declare a variable, you just define the type of data that the variable will store and the variable name in a statement, as follows:

```
datatype variablename;
```

The `datatype` defines what type of data the variable will store. The C programming language uses what's called strict typing, which forces you to declare what type of data each variable will contain. Once you declare the data type for a variable, you can only store that type of data in the variable (more on that in the "Understanding Data Types" section later this hour).

The `variablename` defines the label that you'll use to reference that location in memory from inside your sketch. For example, to declare a variable to hold an integer data type, you use the following:

```
int pin;
```

You have to follow a few rules when declaring C variable names:

- The variable name must contain only letters, numbers, underscores, or the dollar sign.
- The variable name must start with a letter.
- The variable name is case sensitive.
- There is no limit to the variable name length.

Because there is no limit to the length of the variable name, it's a good idea to use meaningful variable names for the data in your sketch. For example, it's better to use variable names like `pin`, `blinkrate`, and `speed` rather than generic names like `a`, `b`, and `c`. That makes troubleshooting code a lot easier, and it reminds you just what each variable does if you have to come back to your sketch several months later.

Another common practice in C coding that's also used by the Arduino developers is a method called camel case. Camel case combines two or more words in a variable name, capitalizing the first letter of each word, with the exception of the first word, as follows:

```
int blinkRateSetting;
```

That makes it a little easier to read and recognize multiword variable names.

As you might expect, each variable name that you declare in your sketch must be unique. You can't declare two variables with the same name; otherwise, you'll get an error message in the IDE console window when you try to compile your sketch.

Defining Variable Values

The second part of the process is to assign a value to the declared variable. To assign a value to a variable, you use the assignment operator, which in the C language is an equal sign:

```
blinkRateSetting = 10;
```

TIP

Terminating a Statement

In the C programming language, you must terminate most statements with a semicolon so that the compiler knows when one statement ends and another starts. We examine which statements don't use the semicolon later this hour.

This is the step that actually assigns the variable to a location in memory. The compiler looks for an available location in memory large enough to hold the data type and stores the assigned value to that location.

You can take a shortcut by declaring the variable and assigning it a value in one statement:

```
int blinkRateSetting = 10;
```

With this method, the compiler assigns the memory location for the variable and stores the data value in one step.

Understanding Data Types

The data type that you declare for a variable defines how large of an area of memory the system needs to reserve to store the data. You can use many different data types in the Arduino programming language. Table 5.1 lists the different data types that Arduino uses to store values, how much memory space they each require, and the maximum values they can hold.

TABLE 5.1 Arduino Data Types

Data Type	Size (Bytes)	Value Range
boolean	1	Logic true or false
char	1	–128 to +127
byte	1	0 to 255
int	2	–32,768 to 32,767
word	2	0 to 65,535
long	4	–2,147,483,648 to 2,147,483,647
float	4	–3.4028235E+38 to 3.4028235E+38
double	4	–3.4028235E+38 to 3.4028235E+38

The `int` and `long` data types store only whole number values. The `float` and `double` data types can store values that contain decimal places, such as 10.5 or –1430.443456.

CAUTION

Integer and Floating-Point Values

Be careful when you're working with numbers in your Arduino sketches. When you use integer values, any mathematical operations you use will result in an integer value. So, dividing 5 by 3 will result in a value of 1. To retain precise results, you have to use floating-point values.

After you define a data type for a variable, that's the only type of data the variable can contain. For example, once you declare a data type as `float`

```
float percent;
```

that defines the storage format the Arduino will use to store the data, and you can't change it. If you store an integer value of 1 in the `percent` variable, the Arduino will still store it in floating-point format as 1.0.

Storing character data (called strings) differs a little bit in the C programming language. When working with string values, the C language converts each character in the string to a binary value using an encoding method, such as the ASCII code, and then stores that value using the `char` data type. It stores each character in the string in sequential order in memory so that it can read the value back in the same order to reproduce the string value. To indicate the end of the string, C places a null character (a 0) as the last byte of the string. This is called a null-terminated string. You'll learn much more about how to use strings in your Arduino sketches later on in Hour 8, "Working with Strings."

Variable Qualifiers

You can also use variable qualifiers to modify how the compiler handles a variable in the sketch. The `const` variable qualifier tells the compiler that the value assigned to the variable won't change. These types of values are called constants.

In the Arduino programming language, it's somewhat common practice to use all uppercase letters for constants:

```
const float TAX = 0.07;
```

The Arduino library contains several different constants that are predefined with values you'll commonly use in your Arduino sketches, such as `HIGH` for the value 1, and `LOW` for the value 0. Constants make reading the sketch code a bit easier to follow, because you can use meaningful constant names instead of obtuse values.

The other type of variable qualifier is the `unsigned` keyword. The `unsigned` keyword tells the compiler to not bother reserving a bit to indicate the sign of the value; all the values stored in that variable will be positive values.

For example, by default, when you define a variable using an `int` data type, the compiler uses 16 bits (2 bytes) to store the value. However, it uses 1 bit to indicate the sign of the value (0 for positive, or 1 for negative), leaving only 15 bits for the actual value. Those 15 bits can only store integer values from 0 to 32,767, so the maximum value range for a signed integer variable is –32,768 to +32,767.

When you apply the `unsigned` qualifier to a variable, it indicates that the value can only be a positive number, so the compiler doesn't need to reserve 1 bit for the sign. This allows the compiler to use all 16 bits to store the value, so the value range can now be from 0 to 65,535.

Table 5.2 shows the unsigned version range of different data types.

TABLE 5.2 Unsigned Data Type Value Ranges

Data Type	Size (Bytes)	Value Range
`unsigned char`	1	0 to 255
`unsigned int`	2	0 to 65,535
`unsigned long`	4	0 to 4,294,967,295

With the `unsigned` qualifier, you have the ability to store some pretty large integer numbers in your programs.

Variable Scope

The last feature of variables that you'll need to know about is variable scope. Variable scope defines where the variable can be used within the Arduino sketch. There are two basic levels of variable scope:

- Local variables
- Global variables

You declare local variables inside a function, and they apply only inside that function code block. For example, if you declare a variable inside the `setup` function, that variable value is available only inside the `setup` function. If you try to use the variable in the `loop` function, you'll get an error message.

In contrast, you can use global variables anywhere in the sketch. It's common practice to declare global variables at the very start of your Arduino sketch, before you define the `setup` function. That makes it easier to see all the global variables that the sketch uses.

Once you declare a global variable, you can use it in either the `setup` or `loop` functions, or any other functions that you create in your sketch. Each time you assign a value to the variable, you can retrieve it from any function.

Using Operators

Just storing data in variables doesn't make for very exciting programs. At some point, you'll want to actually do something with the data you store. This section covers some of the operations that you can perform with your data in the Arduino programming language.

Standard Math Operators

The most basic thing you'll want to do is manipulate numbers, whether it's a counter keeping track of how many loops your sketch takes or a variable that determines how fast the lights in your project should blink. The Arduino library contains all the standard mathematical operators that you're used to using from school; it's just that a few of them may look a little odd.

Table 5.3 lists the different math operators available in the Arduino library.

TABLE 5.3 Arduino Math Operators

Operator	Description
+	Addition
-	Subtraction
*	Multiplication

Operator	Description
/	Division
%	Modulus
++	Increment
--	Decrement
!	Logical NOT
&&	Logical AND
||	Logical OR
&	Bitwise AND
		Bitwise OR
<<	Left shift
>>	Right shift

You should recognize most of these operators from math class. The C programming language also uses the asterisk for multiplication, and the forward slash for division. The modulus operator differs a little; it returns the remainder of the division (what we used to call the "leftover" part).

You'll notice from the table that there are two types of AND and OR operators. There's a subtle difference between the bitwise and logical versions of these operators. You use the bitwise operators in what are called binary calculations. You use binary calculations to perform binary math using binary values.

The logical operators allow you to apply Boolean logic, such as combining values using the AND operation.

To use the math operators in your Arduino programs, you just write your equations in the right side of an assignment statement:

```
blinkRateSetting = 10000 / inputVoltage;
```

The compiler evaluates the mathematical expression on the right side of the assignment statement, and then assigns the result to the variable on the left side.

CAUTION

Assignments Versus Equations

Don't confuse assignment statements with mathematical equations. With assignments, you can use the same variable on both sides of the equal sign:

```
counter = counter + 1;
```

This doesn't make any sense as a mathematical equation, but it is a common assignment statement. The compiler retrieves the current value stored in the `counter` variable, adds 1 to it, and then stores the result back into the `counter` variable location.

Using Compound Operators

Compound operators create shortcuts for simple assignments that you'll commonly use in your Arduino sketches. For example, if you're performing an operation on a variable and plan on storing the result in the same variable, you don't have to use the long format:

```
counter = counter + 1;
```

Instead, you can use the associated addition compound operator:

```
counter += 1;
```

The += compound operator adds the result of the right side equation to the value of the variable you specify on the left side, and stores the result back in the left side variable. This feature works for all the mathematical operators used in the C programming language.

Exploring the Order of Operations

As you might expect, C follows all the standard rules of mathematical calculations, including the order of operations. For example, in the assignment

```
result = 2 + 5 * 5;
```

the compiler first performs the multiplication of 5 times 5, and then adds 2 to that result, resulting in a final value of 27, which is assigned to the `result` variable.

And just like in math, C allows you to change the order of operations using parentheses:

```
result2 = (2 + 5) * 5
```

Now the compiler first adds 2 and 5, and then multiplies the result by 5, resulting in a value of 35.

You can nest parentheses as deep as you need in your calculations. Just be careful to make sure that you match up all the opening and closing parentheses pairs.

Exploring Arduino Functions

The Arduino programming language also contains some standard libraries to provide prebuilt functions for us to use. Functions are somewhat of a black box: You send values to the function, the function performs some type of operation on the values, and then you receive a value back from the function. (To learn more about how to create your own functions, see Hour 10, "Creating Functions.")

This section covers some of the standard Arduino functions that will come in handy as you write your Arduino sketches.

Using Serial Output

Before we get too far, let's talk about a special feature of the Arduino: the ability to output data to an external serial device. This feature provides an easy way for us to monitor what's going on in our Arduino sketches, and will be invaluable to you as you debug your sketches.

To output data to the serial port on the Arduino, you need to use the special `Serial` class, along with its built-in functions. Hour 17, "Communicating with Devices," examines the Serial class in much more detail, but for now we're just interested in three basic functions in the class:

- `Serial.begin()` initializes the serial port for input and output.
- `Serial.print()` outputs a text string to the serial port.
- `Serial.println()` outputs a text string to the serial port and terminates it with a return and linefeed character.

You can use the `Serial.print()` and `Serial.println()` functions in your Arduino sketches to output information from your sketch to view in the serial monitor in the Arduino IDE. This makes for an excellent troubleshooting tool at your disposal.

TRY IT YOURSELF ▼

Using the `Serial` Class Functions to Debug Arduino Sketches

This exercise shows you how to add the `Serial.print()` and `Serial.println()` functions to your Arduino sketch to view variables as your sketch runs. Just follow these steps:

1. Open the Arduino IDE, and enter this code in the editor window:

```
int radius = 0;
float area;

void setup() {
```

```
    Serial.begin(9600);
}

void loop() {
    radius = radius + 1;
    area = radius * radius * 3.14;
    Serial.print("A circle with radius ");
    Serial.print(radius);
    Serial.print(" has an area of ");
    Serial.println(area);
    delay(1000);
}
```

2. Save the sketch as sketch0501.
3. Compile and upload the sketch to your Arduino.
4. Open the serial monitor tool and watch the output.

The output should look something like this:

```
A circle with radius 1 has an area of 3.14
A circle with radius 2 has an area of 12.56
A circle with radius 3 has an area of 28.26
A circle with radius 4 has an area of 50.24
A circle with radius 5 has an area of 78.50
```

The `Serial.print()` function displays a text string, or the value of a variable, but keeps the output cursor on the same line. The `Serial.println()` function displays the text string or variable value, then adds the newline character to start a new line of output.

That's how you can add `Serial.println()` functions to your sketch to watch the values of variables as your sketch is running on the Arduino.

Working with Time

With real-time applications, your sketch will often need to have some knowledge of time. The Arduino library contains four functions that let you access time features in the Arduino:

- **`delay(x)`**: Pauses the sketch for *x* milliseconds.
- **`delayMicroseconds(x)`**: Pauses the sketch for *x* microseconds.
- **`micros()`**: Returns the number of microseconds since the Arduino was reset.
- **`millis()`**: Returns the number of milliseconds since the Arduino was reset.

The `delay()` and `delayMicroseconds()` functions are handy when you want to slow down the output, like we did in the previous example, or if you need for your sketch to wait for a predetermined amount of time before moving on to the next step.

The `millis()` and `micros()` functions enable you to take a peek at how long the Arduino has been running your sketch. Although this is not related to the real time of day, you can use this feature to get a feel for real time spans within your sketch.

Performing Advanced Math

Yes, the basic math operators are useful, but your sketch might sometimes need to use some more advanced mathematical processing. Although not quite as complete as some programming languages, the Arduino programming language does have some support for advanced math functions. Table 5.4 shows what advanced math functions you have to work with in your sketches.

TABLE 5.4 Advanced Arduino Math Functions

Function	Description
abs(*x*)	Returns the absolute value of *x*
constrain(*x*, *a*, *b*)	Returns *x* if *x* is between *a* and *b* (otherwise returns *a* if *x* is lower than *a*, or *b* if *x* is higher than *b*)
cos(*x*)	Returns the cosine of *x* (specified in radians)
map(*x*, *fromLow*, *fromHigh*, *toLow*, *toHigh*)	Remaps the value *x* from the range *fromLow* to *fromHigh* to the range *toLow* to *toHigh*
max(*x*, *y*)	Returns the larger value of *x* or *y*
min(*x*, *y*)	Returns the smaller value of *x* or *y*
pow(*x*, *y*)	Returns the value of *x* raised to the power of *y*
sin(*x*)	Returns the sine of *x* (specified in radians)
sqrt(*x*)	Returns the square root of *x*
tan(*x*)	Returns the tangent of *x* (specified in radians)

You probably recognize most of these functions. The `map()` and `constrain()` functions are a little odd from what you'd see in normal math libraries. They're mostly used when working with sensors. They allow you to keep the values returned by the sensors within a specific range that your sketch can manage.

Generating Random Numbers

When working with sketches, you'll often run into the situation where you need to generate a random number. Most programming languages include a random number generator function, and the Arduino programming language is one of them.

Two functions are available for working with random numbers:

- **`random([min], max)`**: Returns a random number between *min* and *max* - 1. The *min* parameter is optional. If you just specify one parameter, the *min* parameter defaults to 0.
- **`randomSeed(seed)`**: Initializes the random number generator, causing it to restart at an arbitrary point in a random sequence.

With computers, random numbers aren't really all that random. They produce random sequences of repeatable numbers. You may notice if you call the `random()` function enough times that you'll start repeating the same "random" numbers.

To get around that, the `randomSeed()` function allows you to select where in the random sequence the `random()` function starts selecting values. That helps lessen the frequency of repeatable numbers.

Using Bit Manipulation

The bit manipulation group of functions in the Arduino library allows you to work at the bit-level of values. Table 5.5 shows the bit manipulation functions available.

TABLE 5.5 The Arduino Bit Manipulation Functions

Function	Description
bit(*n*)	Returns the value of the specified bit location
bitClear(*x*, *n*)	Clears bit *n* of numeric value *x*
bitRead(*x*, *n*)	Reads bit *n* of numeric value *x*
bitSet(*x*, *n*)	Sets bit *n* of numeric value *x*
bitWrite(*x*, *n*, *b*)	Writes bit *n* of numeric value *x* with value *b*
highByte(*x*)	Returns the high (leftmost) byte of word *x*
lowByte(*x*)	Returns the low (rightmost) byte of word *x*

When working with sensors or multiple inputs, you'll often need to know which bits of a value are set. These functions help you work at the bit level with your data values.

Summary

This hour examined the basics of handling data in Arduino sketches. It walked through how to declare variables in your sketches and how to assign values to them. It also discussed the C language data types and how to apply them to the data in your sketches. You then learned about operators, and how the Arduino language uses built-in math operators to perform standard mathematical operations on data. The discussion then turned to some of the Arduino functions available in the standard library. There are functions for communicating with the serial output on the Arduino, in addition to functions for performing advanced math and time features.

The next hour examines how to control your Arduino programs by testing data conditions and selecting which sections of the sketch code to run.

Workshop

Quiz

1. When does the compiler assign a memory location to store a variable value?
   A. When you declare the variable name
   B. When you define the variable value
   C. When you first run the sketch
   D. When you upload the sketch
2. You can store a floating-point value in a variable that was previously declared as an integer data type. True or false?
3. Which `Serial` function should you use to output a variable value and start a new line in the output?

Answers

1. B. The compiler determines the location to store a variable value in memory when you assign a value to it in the sketch.
2. False. The Arduino programming language uses strict typing of variables. When you declare a variable as an integer data type, the compiler only reserves enough space in memory for that type of data; you can't store another data type in that variable.
3. The `Serial.println()` function allows you to output the value of a variable to the serial port on the Arduino, and sends a newline character after the value so that the serial monitor starts a new line in the output.

Q&A

Q. What is the largest sized floating-point value the Arduino sketch can handle?

A. The floating-point representation of a value in the Arduino uses 4 bytes (or 32 bits) to store the value: 1 bit for the sign, 8 bits for the exponent, and 23 bits for the significand. The maximum value this can store is about 3.4 times 10 to the 38th power.

Q. Can you convert a variable from one data type to another?

A. Yes, sort of. You can use a feature called type casting to tell the Arduino to store a value using a different data type. This is somewhat of a tricky feature of the C programming language and can cause problems if you're not careful when using it.

HOUR 6
Structured Commands

What You'll Learn in This Hour:

- Using `if-then` statements
- Adding `else` sections
- Stringing `if-else` statements together
- Testing conditions
- Using the `switch` statement

In the Arduino sketches we've discussed so far, the Arduino processes each individual statement in the sketch in the order that it appears. This works out fine for sequential operations, where you want all the operations to process in the proper order. However, this isn't how all sketches operate.

Many sketches require some sort of logic flow control between the statements in the sketch. This means that the Arduino executes certain statements given one set of circumstances, but has the ability to execute other statements given a different set of circumstances. There is a whole class of statements that allow the Arduino to skip over or loop through statements based on conditions of variables or values. These statements are generally referred to as structured commands.

Quite a few structured commands are available in the Arduino programming language, so we'll look at them individually. In this chapter, we'll look at the `if` and `switch` structured commands.

Working with the `if` Statement

The most basic type of structured command is the `if` statement. The `if` statement in C has the following basic format:

```
if (condition)
statement
```

You'll often hear of this type of statement referred to as an `if-then` statement. As you can see, though, there's not actually a "then" keyword in the C language version of the statement.

C uses the line orientation to act as the "then" keyword. The Arduino evaluates the condition in the parentheses, and then either executes the statement on the next line if the condition returns a `true` logic value or skips the statement on the next line if the condition returns a `false` logic value.

Here are a few examples to show this in action:

```
int x = 50;
if (x == 50)
   Serial.println("The value is 50");

if (x < 100)
   Serial.println("The value is less than 100");

if (x > 100)
   Serial.println("The value is more than 100");
```

In the first example, the condition checks to see whether the variable `x` is equal to 50. (We talk about the double equal sign in the "Understanding Comparison Conditions" section later in this hour.) Because it is, the compiler executes the `Serial.println()` statement on the next line and prints the string.

Likewise, the second example checks to see whether the value stored in the `x` variable is less than 100. It is, and so the compiler executes the second `Serial.println()` statement to display the string.

However, in the third example, the value stored in the `x` variable is not greater than 100, so the condition returns a `false` logic value, causing the compiler to skip the last `Serial.println()` statement.

Grouping Multiple Statements

The basic `if` statement format allows for you to process one statement based on the outcome of the condition. More often than not, though, you'll want to group multiple statements together, based on the outcome of the condition.

To group a bunch of statements together, you must enclose them within an opening and closing brace, as follows:

```
int x = 50;
if (x == 50) {
    Serial.println("The x variable has been set");
    Serial.println("and the value is 50");
}
```

In this example, the output from both print lines will appear in the serial monitor output. All the statements between the braces are considered part of the "then" section of the statement and are controlled by the condition.

One thing about `if` statements is that it can get tricky trying to pick out the code contained with the braces that's controlled by the `if` condition. One way to help that is to indent the code inside the braces, as shown in the example.

You can either manually format your sketch to look like that, or you can let the Arduino IDE help out with that. Just left-align all the code in the sketch, and then choose Tools > Auto Format from the menu bar. The Arduino IDE will automatically determine which code should be indented.

TRY IT YOURSELF ▼

Working with `if` Statements

To experiment with `if` statements in your Arduino sketches, follow these steps:

1. Open the Arduino IDE, and then enter this code into the editor window:

```
void setup() {
int x = 50;
Serial.begin(9600);
if (x > 25) {
Serial.print("The value if x is:");
Serial.println(x);
Serial.println("This value is greater than 25");
}
Serial.println("This statement executes no matter what the value is");
}

void loop() {
}
```

2. Choose Tools > Auto Format from the menu bar. This will format the code to indent the statements inside the `if` code block.
3. Save the sketch as sketch0601.
4. To verify, compile, and upload the sketch to the Arduino, click the upload icon on the toolbar, or choose File > Upload from the menu bar.
5. Open the serial monitor by clicking the serial monitor icon on the toolbar, or choose Tools > Serial Monitor from the menu bar. This opens the serial monitor application and restarts the Arduino. You should see the following output:

```
The value of x is 50
This value is greater than 25
This statement executes no matter what the value is
```

6. Change the code to set the value of x to 25 in the assignment statement, and then recompile and upload the sketch to the Arduino.

7. Open the serial monitor. You should now see the following output only:

```
This statement executes no matter what the value is
```

Because the new value of x causes the condition to evaluate to a `false` value, the Arduino skips the statements inside the "then" code section, but picks up with the next `Serial.println()` statement that's outside of the braces.

TIP

Restarting a Sketch

Because this sketch just runs once in the `setup()` function then stops, you can rerun the sketch by pressing the Reset button on the Arduino. For the Arduino Uno R3, that button is at the upper-left corner of the circuit board.

Using `else` Statements

In the `if` statement, you have only one option of whether to run (or not) statements. If the condition returns a `false` logic value, the compiler just moves on to the next statement in the sketch. It would be nice to be able to execute an alternative set of statements when the condition is `false`. That's exactly what the `else` statement allows us to do.

The `else` statement provides another group of commands in the `if` statement:

```
int x = 25;
if (x == 50)
       Serial.println("The value is 50");
else
       Serial.println("The value is not 50");
```

The statement after the `else` keyword only processes when the `if` condition is `false`.

Just like with the `if` statement code block, you can use braces to combine multiple statements in the `else` code block:

```
void setup() {
   int x = 25;
   Serial.begin(9600);
   if (x == 50) {
       Serial.println("The x variable has been set");
       Serial.println("and the value is 50");
   } else {
       Serial.println("The x variable has been set");
       Serial.println("and the value is not 50");
   }
   Serial.println("This ends the test");
}

void loop() {
}
```

You can control the output by adjusting the value you assign to the `x` variable. When you run the sketch as is, you'll get this output:

```
The x variable has been set
And the value is not 50
This ends the test
```

If you change the value of `x` to 50, you'll get this output:

```
The x variable has been set
And the value is 50
This ends the test
```

You can also use the Auto Format feature in the Arduino integrated development environment (IDE) to format the `else` code block statements and the `if` code block statements.

Using `else if` Statements

So far, you've seen how to control a block of statements using either the `if` statement or the `if` and `else` statements combination. That gives you quite a bit of flexibility in controlling how your scripts work. However, there's more!

You'll sometimes need to compare a value against multiple ranges of conditions. One way to solve that is to string multiple `if` statements back to back:

```
int x = 45;
if (x > 100)
    Serial.println("The value of x is very large");
if (x > 50)
    Serial.println("The value of x is medium");
```

```
if (x > 25)
    Serial.println("The value of x is small");
if (x <= 25)
    Serial.println("The value of x is very small");
```

With this format, only one of the `System.println()` statements will execute, based on the value stored in the x variable:

```
The value of x is small
```

That works, but it is somewhat of an ugly way to solve the problem. Fortunately, there's an easier solution for us.

The C language allows you to chain `if-else` statements together using the `else if` statement, with a catchall `else` statement at the end. The basic format of the `else if` statement looks like this:

```
if (condition1)
   statement1;
else if (condition2)
   statement2;
else
   statement3;
```

When the Arduino runs this code, it first checks the *`condition1`* result. If that returns a `true` value, the Arduino runs *`statement1`*, and then exits the `if-else if-else` statements.

If *`condition1`* evaluates to a `false` value, the Arduino then checks the *`condition2`* result. If that returns a `true` value, it runs *`statement2`*, and then exits the `if-else if-else` statement.

If *`condition2`* evaluates to a `false` value, it runs *`statement3`*, and then exits the `if-else if-else` statement.

Listing 6.1 shows the sketch0602 code, which is an example of how to use the `else if` statement in a program.

LISTING 6.1 The sketch0602 Code Example

```
void setup() {
int x = 45;
Serial.begin(9600);
if (x > 100)
    Serial.println("The value of x is very large");
else if (x > 50)
    Serial.println("The value of x is medium");
else if (x > 25)
    Serial.println("The value of x is small");
else
```

```
    Serial.println("The value of x is very small");
Serial.println("This is the end of the test");
}

void loop() {
}
```

When you run the sketch0602 code, only one `Serial.println()` statement will execute, based on the value you set the `x` variable to. By default, you'll see this output in the serial monitor:

```
The value of x is small
This is the end of the test
```

This gives you complete control over just what code statements the Arduino runs in the sketch.

Understanding Comparison Conditions

The operation of the `if` statement revolves around the comparisons that you make. The Arduino programming language provides quite a variety of comparison operators that enable you to check all types of data. This section covers the different types of comparisons you have available in your Arduino sketches.

Numeric Comparisons

The most common type of comparisons has to do with comparing numeric values. The Arduino programming language provides a set of operators for performing numeric comparisons in your `if` statement conditions. Table 6.1 shows the numeric comparison operators that the Arduino programming language supports.

TABLE 6.1 Numeric Comparison Operators

Operator	Description
==	Equal
!=	Not equal
<>	Not equal
>	Greater than
>=	Greater than or equal
>	Less than
<=	Less than or equal

The comparison operators return a logical `true` value if the comparison succeeds or a logical false value if the comparison fails. For example, the statement

```
if (x >= y)
  Serial.println("x is larger than y");
```

will execute the `Serial.println()` statement only if the value of the `x` variable is greater than or equal to the value of the `y` variable.

CAUTION

The Equality Comparison Operator

Be careful with the equal comparison. If you accidentally use a single equal sign, that becomes an assignment statement and not a comparison. The Arduino will process the assignment and then exit with a `true` value every time. Most likely that's not what you wanted to do.

TIP

String Comparisons

Because of the odd way the Arduino programming language stores string values, you can't use a standard comparison operator to compare them. Hour 8, "Working with Strings," takes an in-depth look at how to use some Arduino functions to compare string values in your sketches.

Boolean Comparisons

Because the Arduino evaluates the `if` statement condition for a logic value, testing Boolean values is pretty easy:

```
Serial.begin(9600);
boolean x = true;
if (x)
    Serial.println("The value is true");

x = false;
if (x)
    Serial.println("The value is false");
```

Setting a variable value directly to a logical `true` or `false` value is pretty straightforward. However, you can also use Boolean comparisons to test other features of a variable.

If you set a variable to a value, the Arduino will also make a Boolean comparison:

```
int a = 10;
if (a)
    Serial.println("The a variable has been set");
```

However, if a variable contains a value of 0, it will evaluate to a `false` Boolean condition:

```
int testing = 0;
if (testing)
   Serial.println("The testing variable has been set");
```

The comparison for the `testing` variable here will fail because the Arduino equates the 0 assigned to the `testing` variable as a `false` Boolean value. So, be careful when evaluating variables for Boolean values.

TIP

Evaluating Function Results

A related feature to Boolean comparisons is the Arduino's ability to test the result of functions. When you run a function in C, the function returns what's called a return code. You can test the return code using the `if` statement to determine whether the function succeeded or failed.

Creating Compound Conditions

In all the examples so far, we've just used one comparison check within the condition. With the Arduino programming language, you can group multiple comparisons together in a single `if` statement, called a compound condition. This section show some tricks you can use to combine more than one condition check into a single `if` statement.

The Arduino programming language allows you to use the logic operators (see Hour 5, "Learning the Basics of C") to group comparisons together. Because each individual condition check produces a Boolean result value, the Arduino just applies the logic operation to the condition results. The result of the logic operation determines the result of the `if` statement:

```
int a = 1;
int b = 2;
if ((a == 1) && (b == 2))
   Serial.println("Both conditions passed when b=2");

if ((a == 1) && (b == 1))
   Serial.println("Both conditions passed when b=1");
```

When you use the and logic operator, both of the conditions must return a `true` value for the "then" statement to process. If either one fails, the Arduino will skip the "then" code block.

You can also use the `or` logical operator to compound condition checks:

```
if ((a == 1) || (b == 1))
   Serial.println("At least one condition passed");
```

In this situation, if either condition passes, the Arduino will process the "then" statement.

Negating a Condition Check

There's one final `if` statement trick that C programmers like to use. Sometimes when you're writing `if-else` statements, it comes in handy to reverse the order of the "then" and `else` code blocks.

This can be because one of the code blocks is longer than the other, so you want to list the shorter one first, or it may be because the script logic makes more sense to check for a negative condition.

You can negate the result of a condition check by using the logic `not` operator (see Hour 5):

```
int a = 1;
if (!(a == 1))
   Serial.println("The 'a' variable is not equal to 1");

if (!(a == 2))
   Serial.println("The 'a' variable is not equal to 2");
```

The `not` operator reverses the normal result from the equality comparison, so the opposite action occurs from what would have happened without the `not` operator.

TIP

Negating Conditions

You may have noticed that you can negate a condition result by either using the `not` operand or by using the opposite numeric operand (such as a `!=` rather than `==`). Both methods will produce the same result in your Arduino sketch.

Expanding with the `switch` Statement

Often you'll find yourself in a situation where you need to compare a variable against several different possible values. One solution is to write a series of `else if` statements to determine what the variable value is:

```
int test = 2;
if (test == 1)
   Serial.println("The first option was selected");
else if (test == 2)
   Serial.println("The second option was selected");
else if (test == 3)
   Serial.println("The third option was selected");
else
   Serial.println("None of the correct options was selected");
```

The more options there are, the longer this code gets! Instead of writing a long series of `else if` statements, you can use the `switch` statement:

```
int test=3;
switch (test) {
case 1:
   Serial.println("The first option was selected");
   break;
case 2:
   Serial.println("The second option was selected");
   break;
case 3:
   Serial.println("The third option was selected");
   break;
default:
   Serial.println("None of the correct options was selected");
}
```

The `switch` statement uses a standard if-then style condition to evaluate for a result. You then use one or more `case` statements to define possible results from the `switch` condition. The Arduino jumps to the matching case statement in the code, skipping over the other case statements.

However, the Arduino continues to process any code that appears after the `case` statement, including other `case` statements. To avoid this, you can add the `break` statement to the end of the `case` statement code block. That causes the Arduino to jump out of the enter `switch` statement code block.

You can add a `default` statement at the end of the `switch` statement code block. The Arduino jumps to the `default` statement when none of the `case` statements match the result.

The `switch` statement provides a cleaner way of testing a variable for multiple values, without all the overhead of the `if-then-else` statements.

Summary

This hour covered the basics of using the `if` structured command. The `if` statement allows you to set up one or more condition checks on the data you use in your Arduino sketches. You'll find this handy when you need to program any type of logical comparisons in your sketches. The `if` statement by itself allows you to execute one or more statements based on the result of a comparison test. You can add the `else` statement to provide an alternative group of statements to execute if the comparison fails.

You can expand the comparisons by using one or more `else if` statements in the `if` statement. Just continue stringing `else if` statements together to continue comparing additional values.

Finally, you can use the `switch` statement with multiple `case` statements in place of the `else if` statements. That helps make checking multiple values in a variable a bit easier.

The next hour walks through using loops in your Arduino sketches. You can use loops to check multiple sensors using the same code, or you can use them to iterate through data blocks without having to duplicate your code.

Workshop

Quiz

1. What comparison should you use to check if the value stored in the `z` variable is greater than or equal to 10?
   - **A.** `z > 10`
   - **B.** `z < 10`
   - **C.** `z >= 10`
   - **D.** `z == 10`
2. How would you write the `if` statement to display a message only if the value stored in the `z` variable is between 10 and 20 (not including those values)?
3. How would you write `if-else` statements to give a game player status messages if a guess falls within 5, 10, or 15 of the actual value?

Answers

1. C. Don't forget to include the equal sign in the comparison operator when you need to check whether the value is equal to or greater than the desired value. It's easy to forget and just use the greater-than comparator symbol.
2. You could use the following code:

```
if ((z > 10) && (z < 20))
    Serial.println("This is the message")
```

3. You could use the following code:

```
if (z == answer)
   Serial.println("Correct, you guessed the answer!");
else if ((z > (answer - 5)) || (z < (answer + 5)))
   Serial.println("You're within 5 of the answer");
```

```
else if ((z > (answer - 10)) || (z < (answer + 10)))
   Serial.println("You're within 10 of the answer");
else if ((z > (answer - 15)) || (z < (answer + 15)))
   Serial.println("You're within 15 of the answer");
```

Q&A

Q. Is there a limit on how many statements I can place in an `if` or `else` code block?

A. No, you can make the code block as large as needed.

Q. Is there a limit on how many `else if` statements you can place in an `if` statement?

A. No, you can string together as many `else` statements to a single `if` statement as you need.

Q. Is there a limit to how many `case` statements you can place in a `switch` statement?

A. No, you can use as many `case` statements as you need in a single `switch` statement.

Q. Do you have to have a `default` option in a `switch` statement?

A. No. If there isn't a `default` option, and if none of the `case` statements match, no code will be processed in the `switch` statement.

HOUR 7
Programming Loops

What You'll Learn in This Hour:

- Why we need loops
- Exploring the `while` loop
- The `do-while` loop
- Using the `for` loop
- Controlling loops

In Hour 6, "Structured Commands," you saw how to manipulate the flow of an Arduino sketch by checking the values of variables using `if` and `switch` statements. In this hour, we look at some more structured commands that control the flow of your Arduino sketches. You'll learn how to loop through a set of commands until an indicated condition has been met. This hour discusses and demonstrates the `while`, `do-while`, and `for` Arduino looping statements.

Understanding Loops

You've already seen one type of loop used by the Arduino. By default, the Arduino program uses the `loop` function to repeat a block of code statements indefinitely. That allows you to write an application that continues to repeat itself as long as the power to the Arduino is on.

Sometimes, though, in your Arduino sketches you'll find yourself needing to repeat other operations, either until a specific condition has been met, or just repeating a set number of times. An example of this is setting a group of digital ports for input or output mode.

You could just write out all of the `pinMode` function lines individually, but that could get cumbersome:

```
void setup() {
   pinMode(0, INPUT);
   pinMode(1, INPUT);
   pinMode(2, INPUT);
```

```
   pinMode(3, INPUT);
   pinMode(4, INPUT);
   pinMode(5, INPUT);
   pinMode(6, INPUT);
   pinMode(7, INPUT);
   pinMode(8, INPUT);
   pinMode(9, INPUT);
   pinMode(10, INPUT);
   pinMode(11, INPUT);
   pinMode(12, INPUT);
   pinMode(13, INPUT);
}
```

This code would certainly accomplish the task of setting all the digital ports for input, but it sure takes a lot of code to write!

Instead of having to type out each `pinMode` line individually, the Arduino programming language provides a way for you to use a single `pinMode` statement and then run it multiple times for all the lines you want to initialize.

The Arduino programming language provides three types of loop statements to help us simplify repetitive tasks like that:

- The `while` statement
- The `do-while` statement
- The `for` statement

This hour covers each of these statements, plus a couple of other features that come in handy when using loops. First, let's take a look at how to use the `while` statement.

Using `while` Loops

The most basic type of loop is the `while` statement. The `while` statement iterates through a block of code, as long as a specified condition evaluates to a Boolean `true` value. The format of the `while` statement is as follows:

```
while (condition) {
   code statements
}
```

The *`condition`* in the `while` statement uses the exact same comparison operators that the `if-then` statement uses (see Hour 6). The idea is to check the value of a variable that is changed inside the `while` code block. That way your code controls exactly when the loop stops.

Let's run a quick example that demonstrates how the `while` statement works in an Arduino program.

TRY IT YOURSELF 

Experimenting with the `while` Statement

In this example, you create a simple `while` loop to display an output line 10 times. Just follow these steps:

1. Open the Arduino IDE, and then enter this code in the editor window:

```
void setup() {
   Serial.begin(9600);
   int counter = 1;
   while(counter < 11) {
      Serial.print("The counter value is ");
      Serial.println(counter);
      counter = counter + 1;
   }
   Serial.println("The loop has ended");
}
void loop() {
}
```

2. Save the sketch as **sketch0701**.
3. Click the Upload icon on the toolbar to verify, compile, and upload the sketch to your Arduino.
4. Open the serial monitor to view the output of your sketch.

If all goes well, you should see the output shown in Figure 7.1.

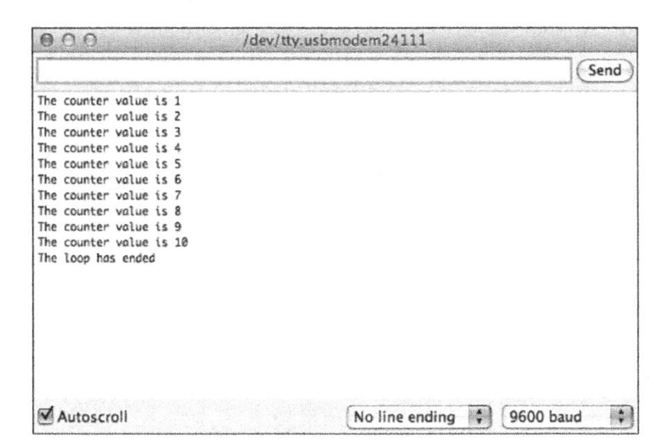

FIGURE 7.1
The output from the sketch0701 sketch.

The loop for the while statement continues as long as the counter variable value is less than 11. Once the counter value is 11, the while statement condition becomes false, so the while loop stops.

WATCH OUT

Endless Loops

Notice that the counter variable value is changed inside the loop. This is a crucial element to using the while statement. If the variable used in the condition doesn't change, your code will get stuck in an endless loop and never exit!

Using do-while Loops

The while statement always checks the condition first, before entering into the loop code block statements. There may be times when you'd like to run the code block statements first, before checking the condition. This is where the do-while loop statement comes in handy.

The format of the do-while statement is as follows:

```
do {
   code statements
} while (condition);
```

When the Arduino runs the do-while statement, it always runs the statements inside the code block first, before evaluating the condition comparison. That means the code is guaranteed to run at least one time, even if the condition is initially false. Let's look at an example of using a do-while statement.

▼ TRY IT YOURSELF

Using the do-while Loop

In this example, you create an Arduino program that uses a do-while statement to loop through a series of statements a set number of times.

1. Open the Arduino IDE, and then enter this code in the editor window:

```
void setup() {v   Serial.begin(9600);
   int counter = 1;
   do {
      Serial.print("The value of the counter is ");
      Serial.println(counter);
      counter = counter + 1;
```

```
    } while (counter < 1);
}

void loop() {
}
```

2. Save the sketch as **sketch0702**.
3. Click the Upload icon on the toolbar to verify, compile, and upload the sketch to your Arduino.
4. Open the serial monitor to view the output of your sketch.

If all goes well, you should see the output shown in Figure 7.2.

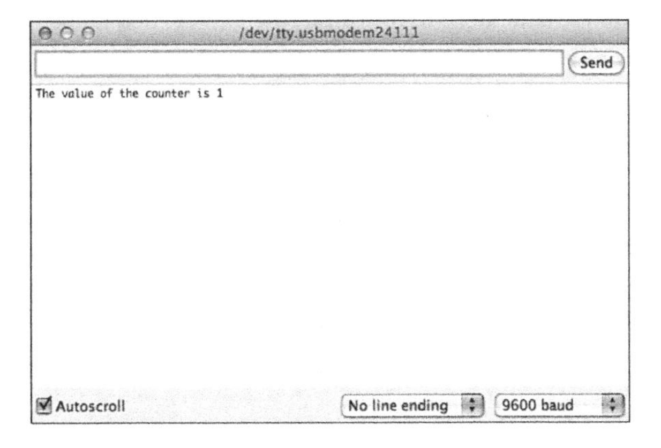

FIGURE 7.2
The output from the sketch0702 sketch.

The condition specified for the `do-while` statement checks to see whether the `counter` variable value is less than 1. Because the assignment statement sets the initial value of the `counter` variable to 1, the very first time the condition is checked, it returns a `false` value. However, the `do-while` loop has already run the statements within the code block before the check, so you'll see output in the serial monitor.

Using `for` Loops

The `while` and `do-while` statements are great ways to iterate through a bunch of data, but they can be a bit cumbersome to use. In both of those statements, you have to make sure that you change a variable value inside the code block so that the loop condition stops when needed.

The Arduino programming language supports an all-in-one type of looping statement called the `for` statement. The `for` statement keeps track of the loop iterations for us automatically.

Here's the basic format of the `for` statement:

```
for(statement1; condition; statement2) {
   code statements
}
```

The first parameter, *`statement1`*, is a C language statement that's run before the loop starts. Normally this statement sets the initial value of a counter used in the loop.

The second parameter, *`condition`*, is the comparison that's evaluated at the start of each loop. As long as the condition evaluates to a `true` value, the `for` loop processes the code statements inside the code block. When the condition evaluates to a `false` value, the Arduino drops out of the loop and continues on in the program code.

The last parameter, *`statement2`*, is a C language statement that's run at the end of each loop. This is normally set to change the value of a counter used in the comparison condition.

A simple example of a `for` statement would look like this:

```
for(i = 0; i < 14; i++) {
   pinMode(i, INPUT);
}
```

These three lines of code just replaced the functionality of having to write out all 14 lines of `pinMode()` statements to set all the digital input lines for input. Now that's handy!

Here's another example of using a `for` statement in an Arduino program.

▼ TRY IT YOURSELF

Using the `for` Statement

Let's use a `for` statement to help simplify setting multiple digital ports for input mode. Just follow these steps:

1. Open the Arduino IDE, and then enter this code in the editor window:

```
void setup() {
   Serial.begin(9600);
   int counter;
   for(counter = 0; counter < 14; counter++) {
      Serial.print("The counter value is ");
      Serial.println(counter);
   }
```

```
    Serial.println("This is the end of the loop");
}

void loop () {
}
```

2. Save the sketch as **sketch0703**.
3. Click the Upload icon to verify, compile, and upload the sketch to your Arduino unit.
4. Open the serial monitor tool to view the sketch output.

You should see the output shown in Figure 7.3.

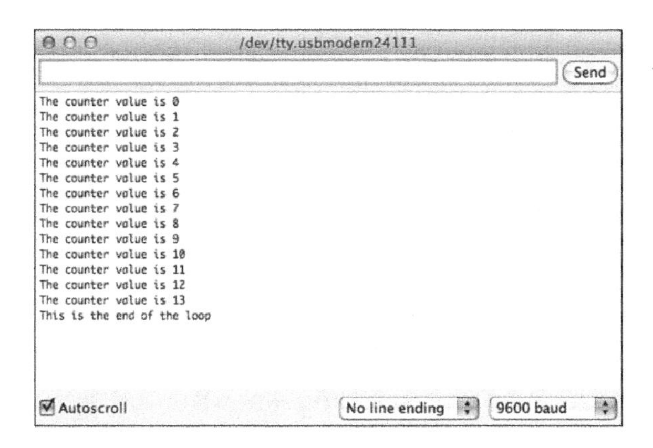

FIGURE 7.3
Output from running the sketch0703 sketch.

Now things are starting to get fancy. You no longer have to keep track of a separate counter variable inside the code block; the `for` statement does all that for you.

Using Arrays in Your Loops

As you can tell, one thing that loops are great at is processing a series of multiple values. However, it won't do you any good to loop through multiple values if all the data is stored in different variables.

Fortunately, the C programming language provides a way to reference multiple values using a single variable name, thus enabling you to easily iterate through the different values in a loop.

This section discusses how to use array variables in your loops to help simplify handling large amounts of data.

Creating Arrays

An array stores multiple data values of the same data type in a block of memory, allowing you to reference the variables using the same variable name. The way it does that is with an index value. The index value points to a specific data value stored in the array.

The format to declare an array variable is as follows:

```
datatype variablename[size];
```

The *datatype* keyword is a standard data type used to declare the variable, such as `int` or `float`. The *size* is a numeric value that indicates how many data values of the specified data type the array will hold. Here's an example of declaring an array of integer values:

```
int myarray[10];
```

The statement declares an array variable called `myarray` that can store up to 10 integer data values in memory. By default, the array data values are empty.

You can reference each data value location by specifying the index with the array variable using square brackets:

```
myarray[0] = 20;
myarray[1] = 30;
myarray[2] = 40;
myarray[3] = 50;
myarray[4] = 100;
```

These statements assign values to the first five data value locations in the `myarray` array variable.

WATCH OUT

Array Indexes

Note that the first data value location in an array is assigned the index value 0.

As with a normal variable, you can declare an array variable and assign it values in a single statement, like this:

```
int myarray[10] = {20, 30, 40, 50, 100};
```

This statement declares the myarray array variable to hold up to 10 integer values, and assigns values to the first 5 data value locations (index values 0 through 4). The braces are required to indicate the values all belong to the same array.

Similarly, to retrieve a specific data value stored in the array variable, you just reference the appropriate index location value:

```
area = width * myarray[0];
```

The Arduino retrieves the data value stored in the specified index location and uses it in the equation for the assignment statement.

Using Loops with Arrays

Besides just using numbers for the array index, you can also use a variable that stores an integer value for the index value in an array:

```
index = 4;
area = width * myarray[index];
```

Now the Arduino first retrieves the value assigned to the index variable, and then it uses that value as the index location for the myarray array variable. It retrieves the value stored at that data value location for the equation in the assignment statement.

Now that you're using a variable as the index, the next step in the process is to change the index value in a loop so that you can iterate through all the data values stored in an array, as shown in this code snippet:

```
int values[5] = {10, 20, 30, 40, 50};
for(int counter = 0; counter < 5; counter++) {
   Serial.print("One value in the array is ");
   Serial.println(values[counter]);
}
```

Note that the for loop counter must start with the value 0, since the array index starts at 0. The condition check in the for loop must also stop before you get to the end of the array; otherwise, the program will return odd values, because it will continue reading memory locations thinking they're part of the array.

Determining the Size of an Array

You may run into situations where you don't know exactly how many data values are in an array variable but you still need to iterate through all of them. This is where the C language sizeof function comes in handy.

The Arduino language `sizeof` function returns the number of bytes used to store an object. You can use it to determine how many bytes an array variable takes in memory, and then with a little math, you can determine just how many data values are currently stored in the array:

```
size = sizeof(myarray) / sizeof(int);
```

The `sizeof(int)` returns the number of bytes the system uses to store an integer data type. By dividing the total size of the array by the size of a single integer value, you can determine just how many data elements are in the array. For example, in the previous `for` loop example, you could use the following:

```
for(counter = 0; counter < (sizeof(value)/sizeof(int)); counter++) {
```

The Arduino will only iterate through the number of data elements defined for the array. This is a common practice in the C programming world, and can save you lots of calculations in your Arduino code.

Using Multiple Variables

Another trick often used in `for` statements is the ability to track multiple counters in a single statement. Instead of initializing just a single counter variable in the `for` statement, you can initialize multiple variables, separated with a comma. Likewise, you can change the values of all those variables at the end of the loop. The format to do that looks like this:

```
int a,b;
for (a = 0, b = 0; a < 10; a++, b++) {
   Serial.print("One value is ");
   Serial.print(a);
   Serial.print(" and the other value is ");
   Serial.println(b);
}
```

This `for` statement uses two counters: the `a` variable and the `b` variable. At the end of each iteration, the program increments both variables, but the condition only checks the value of the `a` variable to determine when to stop the loop.

Nesting Loops

Another popular use of loops is called nesting. Nesting loops is when you place one loop inside another loop, each one controlled by a separate variable.

The trick to using inner loops is that you must complete the inner loop before you complete the outer loop. The closing bracket for the inner loop must be contained within the outer loop code:

```
int a,b;
for(a = 0; a < 10; a++) {
   for(b = 0; b < 10; b++) {
      Serial.print("One value is ");
      Serial.print(a);
      Serial.print(" and the other value is ");
      Serial.println(b);
   }  // closing brace for the 'b' loop
} // closing brace for the 'a' loop
```

If you forget to close out the inner loop, you won't get the results that you planned.

Controlling Loops

Once you start a loop, it will usually continue until the specified condition check becomes false. You can change that behavior using two different types of statements:

- The `break` statement
- The `continue` statement

Let's take a look at how each of these statements works.

The `break` Statement

You use the `break` statement when you need to break out of a loop before the condition would normally stop the loop. Let's take a look at an example of how this works.

TRY IT YOURSELF ▼

Using the `break` Statement

The `break` statement allows you to "jump out" of a loop before it would normally terminate. To test that, run through this demo:

1. Open the Arduino IDE, and then enter this code in the editor window:

```
void setup() {
   Serial.begin(9600);
   int i;
   for(i = 0; i <= 20; i++) {
      if (i == 15)
         break;
```

```
        Serial.print("Currently on iteration: ");
        Serial.println(i);
    }
    Serial.println("This is the end of the test");
}
void loop() {
}
```

2. Save the sketch as **sketch0704**.
3. Click the Upload icon to verify, compile, and upload the sketch to your Arduino unit.
4. Open the serial monitor tool to view the sketch output.

You should see the output shown in Figure 7.4.

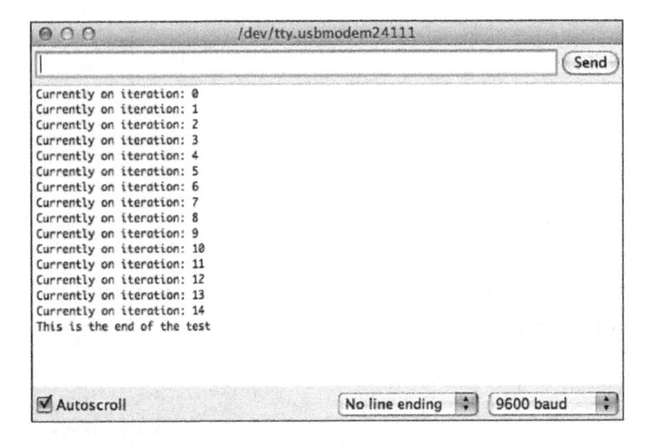

FIGURE 7.4
Output from running the sktech0704 sketch.

The `break` statement terminates the `for` loop in iteration 15, before the defined condition for the loop was met.

The `continue` Statement

The `continue` statement may be a little odd to follow. Instead of telling the Arduino to jump out of a loop, it tells the Arduino to stop processing code inside the loop, but still jumps back to the start of the loop. That might sound somewhat confusing, as you'd think to stop processing inside the loop you'd exit the loop (like the `break` statement does).

Perhaps the easiest way to follow how the `continue` statement works is to watch it in action:

TRY IT YOURSELF ▼

Using the `continue` Statement

In this example, you test to see how the `continue` statement changes the behavior inside a standard `for` loop. Just follow these steps to run the experiment:

1. Open the Arduino IDE, and then enter this code in the editor window:

```
void setup() {
  Serial.begin(9600);
  int i;
  for(i = 0; i < 20; i++) {
    if ((i > 5) && (i < 10)) {
      continue;
    }
    Serial.print("The value of the counter is ");
    Serial.println(i);
  }
  Serial.println("This is the end of the test");
}
void loop() {
}
```

2. Save the sketch as **sketch0705**.
3. Click the Upload icon to verify, compile, and upload the sketch to your Arduino unit.
4. Open the serial monitor tool to view the sketch output.

You should see the output shown in Figure 7.5.

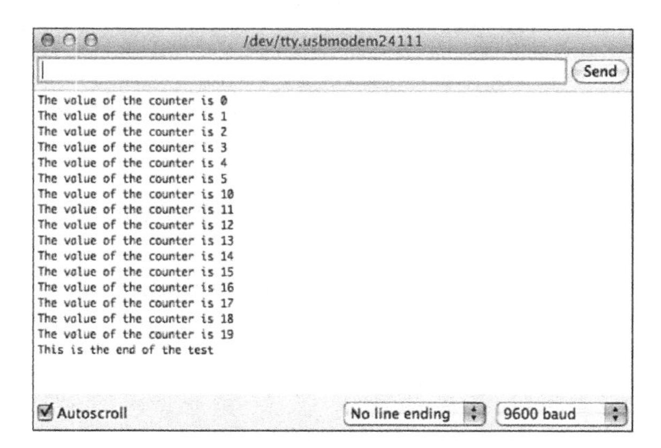

FIGURE 7.5
Output from running the sketch0705 sketch.

The `continue` statement causes the Arduino to stop processing the code inside the loop if the `i` variable value is between 5 and 10, but it returns back to the start of the loop and continues on with the loop iterations. The output in Figure 7.5 shows that the `for` loop continues, even when the `continue` statement processes.

Summary

This hour covered how to iterate through blocks of code multiple times. It's common to run into situations where you need to repeat one or more statements multiple times with different values. By using loop statements, you can reduce the number of statements you have to write. The `while` loop enables you to check a condition that determines when the loop stops. With the `while` loop, it's your responsibility to make sure that the condition value changes inside the loop. The `do-while` loop works similarly to the `while` loop, but it checks the condition at the end of the loop iteration rather than at the start. The `for` loop helps simplify things for us by providing the counter, condition check, and the statement that modifies the counter all in one code statement.

This hour also introduced you to the idea of array variables, which allow you to store multiple data values referenced by the same variable name but with different indexes. These come in handy when working with loops.

In the next hour, you learn how to work with text in your Arduino programs, which will prove handy if you're using any type of human interface for your sketches.

Workshop

Quiz

1. Which loop type will always run at least one iteration of the loop code block no matter if the condition is met?
   - **A.** The `for` loop
   - **B.** The `while` loop
   - **C.** The `do-while` loop
   - **D.** The `if-then` loop
2. With array variables, you can store multiple data values in the array and then iterate through them using a loop. True or false?
3. How could you write a loop to iterate through values 0 through 20, skipping over values 5 through 8 and 10 through 13?

Answers

1. The do-while loop runs the loop code block first, before checking the condition, so it will always run at least one iteration of the loop code block.

2. True. The array stores multiple data values in a common location in memory, and then allows you to use a numeric index value to reference each individual stored value. You can use a loop to iterate through the index values.

3. You can use the continue statement to skip over the value ranges:

```
int counter;
for(counter = 0; counter <= 20; counter++) {
   if ((counter >= 5) && (counter <= 8))
      continue;
   if ((counter >= 10) && (counter <= 13))
      continue;
   Serial.print("The value if the counter is ");
   Serial.println(counter);
}
```

Q&A

Q. What happens if I make a mistake and define a condition for a loop that is never met?

A. Because the condition will never evaluate to a true value, the loop will never end. This is called an infinite loop and will cause your Arduino to get stuck in the loop forever.

Q. Is it considered bad form to use break statements in loops?

A. Some programmers believe that you should never use break statements in loops and that you should try to write your loops to avoid using the break statement. However, the break statement is there for a reason, and you may run into situations where it makes perfectly good sense to use it.

Q. Is there a limit to how deep I can nest loops in the Arduino code?

A. No, you can have as many nested loops as you need; however, things do start to get somewhat confusing after a couple of layers of loops, so be careful if you go deeper than that.

HOUR 8
Working with Strings

What You'll Learn in This Hour:

- What are strings?
- Handling strings with C
- The Arduino way to handle strings
- Working with strings in your programs

One of the most confusing parts of the C programming language is working with text. Unfortunately, the C language uses a complex way of handling text, which can be somewhat confusing to beginner programmers. Fortunately, the Arduino developers realized that, and helped by creating some built-in string-handling features for the Arduino programming language. This hour first takes a look at how the C language handles strings, and then demonstrates how it's much easier to work with strings in the Arduino environment.

What's a String?

Whereas computers only understand and work with numbers, humans need words to communicate ideas. The trick to interfacing computer programs with humans is to force the computer to produce some type of output that humans can understand.

Any time you need your program to interface with humans, you need to use some type of text output. To do that with a computer program requires the use of the character data type.

The character data type stores letters, numbers, and symbols as a numeric value. The trick comes in using a standard way of mapping numeric values to language characters. One of the more popular character mapping standards for the English-speaking world is the ASCII format.

With ASCII, each character in the English alphabet, each numeric digit, and a handful of special characters, are each assigned a numeric value. The computer can interpret the ASCII number representation of the character to display the proper text based on the value stored in memory.

To store a character in memory, you use the `char` data type:

```
char letter = 'a';
```

When you use this statement, the Arduino doesn't store the actual letter *a* in memory; it stores a numeric representation of the letter *a*. When you retrieve the `letter` variable and display it using the `Serial.print` function, the Arduino retrieves the numeric value, converts it to the letter *a*, and then displays it.

The next step in the process is storing words and sentences. A string value is a series of characters put together to create a word or sentence. The Arduino stores the multiple characters required to create the word or sentence consecutively in memory, then retrieves them in the same order they were stored.

The problem is that the Arduino needs to know just when a string of characters that make up the word or sentence stored in memory ends. Remember, the individual characters are stored as just numbers, so the Arduino processor has no idea of what number in memory represents a character and what doesn't.

The solution is null-terminated strings. In the C programming language, when a string value is stored in memory, a `0` value is placed in the memory location at the end of the characters in the string value (this is called a null value). That way, the Arduino processor knows that when it gets to the null value, the string value is complete.

The next section shows just how to store and work with string values using the standard C programming language.

Understanding C-Style Strings

As you saw in the preceding section, the trick to creating strings in the C programming language is to store a series of characters in memory and then handle them as a single value. In Hour 7, "Programming Loops," you were introduced to the idea of arrays. You can use character arrays to create and work with strings in the C programming language.

To refresh your memory, an array is a series of values of the same data type stored in memory, all associated with the same variable name. You can reference an individual value in the array by specifying its index value in square brackets in the variable name:

```
myarray[0] = 10;
```

The examples in Hour 7 only used numeric values in arrays, but you can use the same method with character values.

This section shows you how to use the C programming method to create and work with strings in your Arduino programs.

Creating Character Strings

A string is nothing more than an array of character values, with the last array data value being a `0` to indicate the null terminator. You can create a string character array using several different formats. To define a simple string, you just create an array with the individual letters of the string, with a null character as the last array value:

```
char test[7] = {'A', ' ', 't', 'e', 's', 't', '\0'};
```

Notice that each character (including the space) is defined as a separate data element in the array. Also, note that the array size that you define must be large enough to hold all the characters in the string plus the terminating null character (represented using the `\0` symbol). To retrieve the stored string value, you just reference it by its array variable name:

```
Serial.println(test);
```

Although this works, it is somewhat of a hassle to have to spell out each character in the string as a separate data element in the array. Fortunately, the C language provides a shortcut:

```
char test[15] = "This is a test";
```

The C compiler knows to create an array out of the characters listed within the double quotes and automatically adds the terminating null character at the end of the array. However, just as with the long format, you must declare the size of the array to be large enough to hold all the characters plus the null character.

Some C compilers (including the one used in the Arduino) allow you to initialize a string array without defining the size:

```
char test[] = "Testing the Arduino string";
```

The compiler automatically reserves enough space in memory to hold the declared string characters and the terminating null character.

BY THE WAY

Strings, Characters, and Quotes

You'll notice in the examples that I used single quotes around the individual characters in the array, but double quotes around the full string value. That's a requirement for the C language, so be careful when defining character and string values!

When you initialize the string, you can make the array size larger than the initial string value:

```
char test[20] = "This is a test";
```

That allows you to change the string value stored in the variable later on in your sketch code to something larger.

WATCH OUT

Overflowing String Values

You must take great care when working with character arrays. The C compiler will allow you to store a value larger than the defined array size without generating an error message. However, the extra characters will "overflow" into the memory location for other stored variables, causing interesting issues with your sketch.

Finally, you can declare a character array without initializing it with a value:

```
char test3[20];
```

This reserves 20 bytes of memory for the string value storage. The downside to just declaring a character array is that you can't just assign a character array value to it. Unfortunately, the C compiler can't determine how to store the string value in the reserved character array. However, some common functions available in the C language provide ways for you to manipulate existing character arrays. The next section shows how to do that.

Working with Character Arrays

After you've created a character array in memory, you'll most likely want to be able to use it in your Arduino programs. To reference the string as a whole, you can just use the character array variable name. Here's an example of how to do that.

▼ TRY IT YOURSELF

Displaying a String

To display stored string values in your Arduino sketches, you can use the standard `Serial.println()` function. Follow these steps to test that out:

1. Open the Arduino IDE, and then enter this code in the editor window:

```
void setup() {
   Serial.begin(9600);
   char msg1[20] = "This is string 1";
   char msg2[20] = "This is string 2";
   Serial.println(msg1);
   Serial.println(msg2);
}
void loop() {
}
```

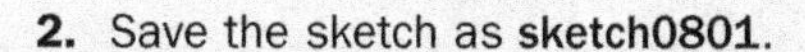

2. Save the sketch as **sketch0801**.
3. Click the Upload icon to verify, compile, and upload the sketch to the Arduino unit
4. Open the serial monitor tool to view the output from your sketch.

The output of the sketch should show the two strings, each string on a separate line, as shown in Figure 8.1.

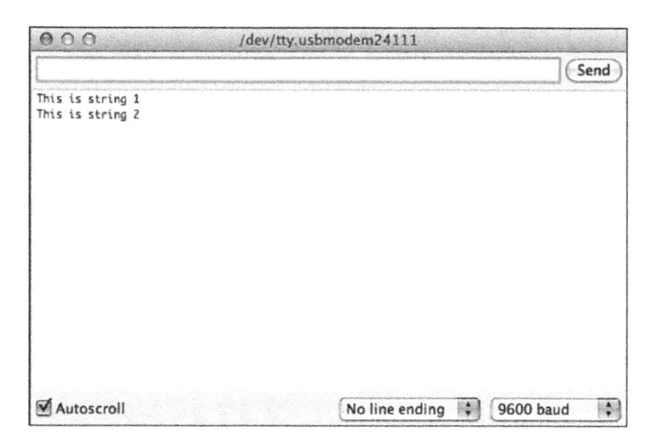

FIGURE 8.1
Output from running the sketch0801 code.

Once you declare the character array for the string and assign it a value, you cannot use standard assignment statements to change the character array like you would with numeric variables.

Instead, you have to use some of the built-in C functions designed for manipulating string values. Table 8.1 lists the string functions that the Arduino programming language supports.

TABLE 8.1 The Arduino String Functions

Function	Description
`strcpy(s1, s2)`	Copies the string *s2* into the *s1* location.
`strcmp(s1, s2)`	Compares string *s1* to *s2*. Returns 0 if they match, a negative value if *s1* > *s2*, or a positive value if *s2* > *s1*.
`strlen(s1)`	Returns the length of the string value.
`strstr(s1, s2)`	Returns the location of substring *s2* in *s1*.

Here's an example of using the strcpy string function to change the value stored in a string variable in an Arduino sketch.

▼ TRY IT YOURSELF

Working with String Functions

You can manipulate string values stored in memory within your sketch code using the standard C language functions. Follow these steps to run this example:

1. Open the Arduino IDE, and then enter this code into the editor window:

```
void setup() {
   Serial.begin(9600);
   char msg[50];
   int rand;
   for(int i = 0; i < 10; i++) {
     rand = random(100);
     if (rand < 50)
       strcpy(msg, " the value is small");
     else
       strcpy(msg, " the value is large");
     Serial.print("The random value is ");
     Serial.print(rand);
     Serial.println(msg);
   }
}
void loop() {
}
```

2. Save the sketch as **sketch0802**.
3. Click the Upload icon to verify, compile, and upload the sketch to the Arduino unit
4. Open the serial monitor tool to view the output from your sketch.

The output of the sketch should show the two strings, each string on a separate line, as shown in Figure 8.2.

The sketch0802 sketch uses the random function to return a random number between 0 and 100. The if-then statement checks to see whether the number is less than 50, and uses the strcpy function to copy a message into the msg variable. This shows that you can dynamically change the value stored in a character array in your sketch, which can come in handy.

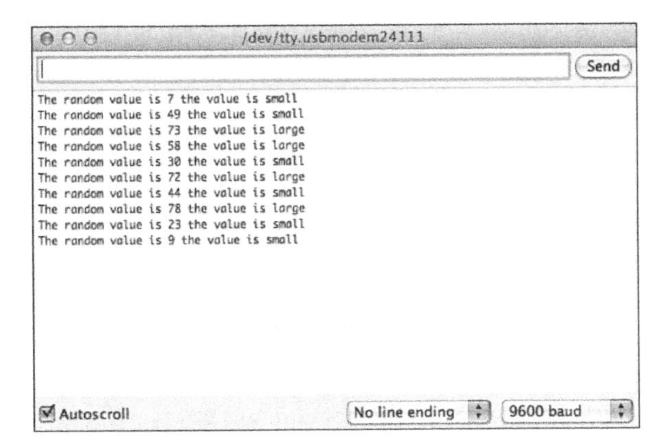

FIGURE 8.2
Output from running the sketch0802 code.

WATCH OUT

Copying Strings

Be careful when using the `strcpy` function to copy string values into character arrays. It's important to remember about the character array size limitation. If you try to copy too large of a string value, you'll overflow the array area in memory.

Comparing Character Arrays

One of the more popular functions you'll need to do with character arrays is to compare them. You do that using the `strcmp` function. How you use it is a little odd, though, so let's work through an example of using it in your Arduino sketches.

The format of the `strcmp` function is fairly simple:

```
strcmp(string1, string2)
```

What's tricky about the `strcmp` function is the value that it returns:

- A `0` if the two string values are equal
- A negative number if *`string1`* is less than *`string2`*
- A positive number if *`string1`* is greater than *`string2`*

A common mistake made by beginning C programmers is to write something like this:

```
if (strcmp(string1, string2))  // the wrong way to use the strcmp function
```

The problem with this statement is that if the two string values are equal, the `strcmp` function returns a `0` value, which the `if-then` statement interprets as a `false` value! If you want to compare two character arrays for equality, you want to use the following:

```
if (strcmp(string1, string2) == 0)
```

Now the Arduino will run the "then" code block if the two string values are equal.

Introducing the Arduino `String` Object

Working with the C language character arrays to handle text in your sketches is not for the faint of heart. Quite a lot of things can (and often do) go wrong if you're not careful.

Fortunately, the Arduino developers took pity on the novice programmers that Arduino users often are and worked hard to create an easier way to create and work with strings in our Arduino sketches. Instead of using an array of character values to handle strings, the Arduino developers created a robust class object that does most of the hard work for us behind the scenes. This section shows you how to use the `String` object to make working with strings much easier.

The `String` Class

The Arduino `String` class allows us to easily create and work with just about any type of string value without the hassles involved with using C-style character arrays.

The easiest way to create a `String` object is to declare a variable using the `String` data type and assign it a value:

```
String test1 = "This is a test string";
```

The `String` object also has a constructor method named `String` that can convert just about any data type value into a `String` object:

```
String test2 = String('a');
String test3 = String("This is a string of characters");
String test4 = String(31);
String test5 = String(20, HEX);
```

The result of all these declaration statements is a `String` object. The last example defines the number base of the value used as the first parameter.

One nice feature of the `String` class is that unlike character arrays, you can use it in normal assignment statements:

```
String string1 = "test";
string1 = "a new test";
```

The value stored in the `string1` object changes in the assignment statement without you having to worry about string overflow; the Arduino takes care of that for you. Let's take a look at an example of using the `String` object in an Arduino sketch.

TRY IT YOURSELF ▼

Manipulating String Objects

In this example, you use the `String` object to manipulate a string value within your Arduino sketch. Just follow these steps:

1. Open the Arduino IDE, and then enter this code into the editor window:

```
1:  void setup() {
2:        Serial.begin(9600);
3:        String string1 = "This is a test";
4:        String string2 = String("This is a second string");
5:        String string3;
6:        Serial.println(string1);
7:        string1 = "This is a longer string value";
8:        Serial.println(string1);
9:        string1 = String(string1 + " with more text");
10:       Serial.println(string1);
11:       Serial.println(string2);
12:       string3 = String(string1 + string2);
13:       Serial.println(string3);
14: }
15:
16: void loop() {
17: }
```

2. Save the code as **sketch0803**.
3. Click the Upload icon to verify, compile, and upload the code to your Arduino.
4. Open the serial monitor to view the output of your sketch.

The sketch creates three `String` objects and manipulates the values in them. Notice in line 7 that the code replaces the value originally stored in the `string1` variable with a longer string value without causing any overflow problems. Lines 9 and 12 demonstrate how to concatenate multiple string values into a single value. When you open the serial monitor, you should see the output shown in Figure 8.3.

Once you declare a `String` object, you can use many different handy methods to manipulate those objects. The next section shows how to do that.

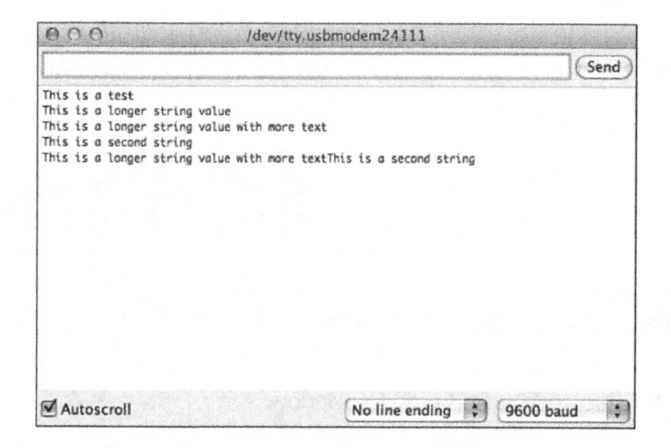

FIGURE 8.3
The output from the sketch0803 code.

Various `String` Object Methods

After you create a `String` object, you can use several different methods to retrieve information about the stored `String` object. Table 8.2 shows these methods.

TABLE 8.2 The `String` Object Methods

Method	Description
`charAt(n)`	Returns the character at the *n*th position in the string.
`compareTo(string2)`	Returns `0` if the string is equal to *`string2`*, a negative number if the string is less than *`string2`*, or a positive number if the string is greater than *`string2`*.
`concat(string1, string2)`	Appends the *`string2`* value to the end of *`string1`*, and creates a new string value.
`endsWith(string2)`	Returns `true` if the string ends with the *`string2`* value.
`equals(string2)`	Returns `true` if the string is equal to *`string2`*.
`equalsIgnoreCase(string2)`	Returns `true` if the string is equal to *`string2`*, ignoring character case.
`getBytes(buf, len)`	Copies *`len`* string characters into the *`buf`* variable.
`indexOf(val, [,from])`	Returns the index location where the string *`val`* starts in the string. By default, it starts at index 0, or you can specify a starting location using the *`from`* parameter. Returns `-1` if *`val`* is not found in the string.

Method	Description
lastIndexOf(*val* [, *from*])	Returns the index location where the string *val* starts in a string. By default, it starts at the end of the string, working toward the front of the string, or you can specify a starting location using the *from* parameter. Returns -1 if *val* is not found in the string.
length()	Returns the number of characters in the string (not counting the terminating null character).
replace(*substring1*, *substring2*)	Returns a new string with the *substring1* value with *substring2* in the original string value.
reserve(*n*)	Reserves a space of *n* characters in memory for a string value.
startsWith(*string2*)	Returns true if the string starts with *string2*.
substring(*from* [, *to*])	Returns a substring of the original string value, starting at the *from* index location. By default, it returns the rest of the string from that location, or you can specify the *to* index value.
toCharArray(*buf*, *len*)	Copies *len* characters in the string to the character array variable *buf*.
toInt()	Returns an integer value created from the string value. The string must start with a numeric character, and the conversion stops at the first non-numeric character in the string.

The String object methods make working with strings a breeze. Remember the ugly strcmp function we had to use with the character array method? The equals function solves that silliness. Here's an example of how to use that in your sketches:

```
String string1 = "This is a test";
String string2 = "this is a test";
if (string1.equals(string2))
   Serial.println("The strings are exactly equal");
if (string1.equalsIgnoreCase(string2))
   Serial.println("The strings are the same without regard to case");
```

If you run this code in your sketch, the first if-then condition check will fail, because the two string values aren't exactly the same. However, using the equalsIgnoreCase function allows us to compare the string values without regard to case. That comes in handy if you need to check answers entered by users, such as yes/no responses to queries.

Manipulating String Objects

Besides the string methods that return answers based on the string value, a handful of methods actually alter the existing string value stored in the `String` object. Table 8.3 shows the different `String` class methods you can use to manipulate a `String` object.

TABLE 8.3 The `String` Object Manipulation Methods

Method	Description
`setCharAt(index, c)`	Replaces the character at *index* with the character *c*.
`toLowerCase()`	Converts the string value to all lowercase letters.
`toUpperCase()`	Converts the string value to all uppercase letters.
`trim()`	Removes any leading and trailing space or tab characters from the string value.

All of these functions enable you to manipulate the string value stored in a variable by changing the characters stored in the object.

WATCH OUT

Replacing String Values

Be careful with the string manipulation methods, because they replace the original string value with the result. The original string value will be lost in the process.

To get an idea of just how this works, let's use the `toUpperCase` method in an example.

▼ TRY IT YOURSELF

Manipulating a String Value

This example changes the value of a string to convert the characters to all uppercase. Just follow these steps:

1. Open the Arduino IDE, and then enter this code into the editor window:

```
void setup() {
   Serial.begin(9600);
   String string1 = "This Is A Mixed Case String Value";
   Serial.println("Before the conversion:");
   Serial.println(string1);
```

```
    string1.toUpperCase();
    Serial.println("After the conversion:");
    Serial.println(string1);
}
void loop() {
}
```

2. Save the file as **sketch0804**.
3. Click the Upload icon to verify, compile, and upload the sketch to your Arduino.
4. Open the serial monitor to restart the sketch and view the output.

You should see the output shown in Figure 8.4 in your serial monitor window.

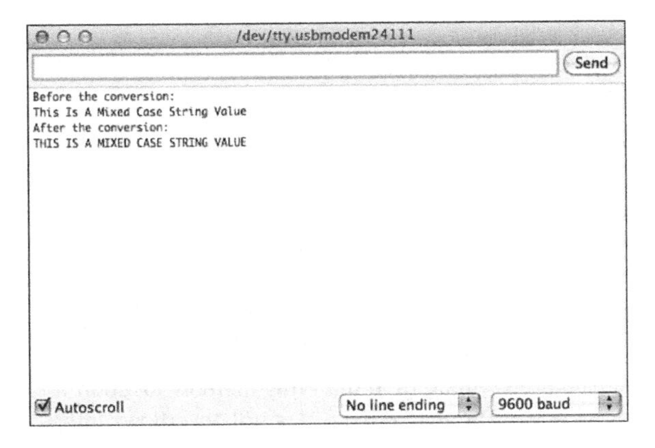

FIGURE 8.4
Output from the sketch0804 code.

This example demonstrates how you can manipulate a `String` object value using the methods in the `String` object class.

Summary

This hour explored the world of using strings in your Arduino sketches. The C style of strings requires that you create a character array to store the individual characters in the string, terminated with a null character. This method can be cumbersome to work with, and can cause problems if you're not careful. You then learned how to use the Arduino `String` object to create and work with string values. With `String` objects, you don't have to worry about the string size, plus there are many more functions that allow you to easily compare and manipulate string values, which makes working with text a lot easier.

The next hour delves into the world of data structures. Sometimes you'll want to be able to group data elements together as a single object, such as for storing data. Data structures allow you to do just that.

Workshop

Quiz

1. Which `String` class method should you use to emulate the `strcmp` character array function?

  A. `compareTo`

  B. `equals`

  C. `equalsIgnoreCase`

  D. `indexOf`

2. The Arduino compiler won't allow you to store more characters in a character array than the size of the array you defined. True or false?

3. How can you check whether a user answers YES or yes to a question?

Answers

1. The `compareTo` method returns a `0` if the two string values match, a negative value if the string is smaller than the compared value, and a positive value if the string is larger than the compared value. This behavior emulates the `strcmp` function used for character arrays.

2. False. The compiler will allow you to store more characters in the array memory location than the size you defined for the character array. The extra characters will "overflow" into the memory location for other variables, causing problems in your programs, so be careful.

3. Use the `equalsIgnoreCase` method to compare the answer string to a yes string value.

Q&A

Q. Does it matter whether I use a character array or a `String` object in my sketches?

A. For functional purposes, no; both methods will work the same. However, `String` objects do take up more memory space than character arrays. So if you're writing a large sketch and are tight on memory, you might want to use character arrays.

Q. Can I create an array of character arrays and then access each string value separately?

A. Yes, but that requires a feature called pointers, which is discussed in Hour 11, "Pointing to Data."

HOUR 9
Implementing Data Structures

What You'll Learn in This Hour:

- What data structures are
- How to create a data structure in C
- How to use data structures in your Arduino sketches
- Using data unions in your sketches

Data is the core of every Arduino sketch. Whether you're monitoring values from analog or digital sensors, or sending data out a network interface, data management can be one of the biggest chores of writing sketches. The Arduino programming language provides some level of data management features that help you organize the data that you use in your sketches. In this hour, we explore two different data management features: structures and unions.

What's a Data Structure?

In Hour 5, "Learning the Basics of C," you learned how to use variables to store data in your Arduino sketches. Each variable stores one data value, of a specific data type, in a predefined location in memory. Although variables are useful, sometimes working with individual variables can get a bit complicated with large amounts of data.

Suppose, for example, that you want to track the data from both an indoor and outdoor temperature sensors at various times in the day. Using variables, you'd have to create a separate variable for each temperature value at each time of the day you want to use:

```
int morningIndoorTemp;
int morningOutdoorTemp;
int noonIndoorTemp;
int noonOutdoorTemp;
int eveningIndoorTemp;
int eveningOutdoorTemp;
```

The more time samples you want to store, the more data variables you have to create. You can see that it won't take long before the number of variables your sketch has to work with starts getting out of hand.

In Hour 7, "Programming Loops," you learned how to create array variables. That helps some, because you can at least use the same variable name to store multiple data values. With array variables, you can store the related temperature samples in the same array:

```
int morningTemps[2];
int noonTemps[2];
int eveningTemps[2];
```

That helps cut down on the number of variables you have to work with from six to three, but now you have to keep track of which array index represents which temperature sensor.

Data structures allow us to define custom data types that group related data elements together into a single object. Using a data structure, you can group the multiple sensor values together into a single data object that your program can handle as a single entity. Instead of referencing the individual data elements using an index, you can use variable names to make it obvious which values are which. This proves when storing and retrieving data in databases, files, or just in memory.

The Arduino handles data structures as a single block in memory. When you create a data structure, the Arduino reserves space in memory based on all the data elements contained in the data structure. Creating, storing, and reading data structures is a simple operation that takes minimal coding effort. The next section demonstrates how to start working with data structures.

Creating Data Structures

Before you can use a data structure in your sketch, you need to define it. To define a data structure in the Arduino, you use the `struct` statement. The generic format for the `struct` statement is as follows:

```
struct name {
   variable list
};
```

The *`name`* defines a unique name used to reference the data structure in your sketch. The *`variable list`* is a set of variable declarations to define the variables contained within the data structure.

Because the data structure definition is part of the variable declarations, you must place it at the top of your Arduino sketch, either before you define the `setup` function or inside the `setup` function. Here's an example of a simple data structure definition:

```
struct sensorinfo {
   char date[9];
   int indoortemp;
   int outdoortemp;
};
```

This data structure definition declares three variables as part of the structure:

- A character array variable to store a date value
- An integer variable to store the indoor temperature from the sensors
- A second integer variable to store the outdoor temperature from the sensor

Notice that you can mix and match the data types stored in a data structure. You don't have to use all the same data types in the structure.

This `struct` statement just defines the data structure format; it doesn't actually create an instance of the data structure for storing data. To do that, you need to declare individual variables using the data structure data type:

```
struct name variable;
```

This statement declares a variable that uses the data structure assigned to the *`name`* structure that you defined earlier. For example:

```
struct sensorinfo morningTemps;
struct sensorinfo noonTemps;
struct sensorinfo eveningtemps;
```

These three statements declare three separate instances of the `sensorinfo` data structure: `morningTemps`, `noonTemps`, and `eveningTemps`. All the variables contain all the data elements defined for the `sensorinfo` data structure and store separate values for those elements.

You can use a shortcut format to declare your sketches in one statement:

```
struct sensorinfo {
   char date[9];
   int indoortemp;
   int outdoortemp;
} morningTemps, noonTemps, eveningTemps;
```

This defines the structure and assigns it to variables all at one time.

BY THE WAY

Unnamed Data Structures

You can also define a data structure and declare variables that use the data structure without assigning a name to the data structure:

```
struct {
   char date[9];
   int indoortemp;
   int outdoortemp;
} morningTemps, noonTemps, eveningTemps;
```

However, if you do that, because there isn't a data structure name associated with the structure, you can't declare any other variables in the sketch with that data structure.

After you declare the data structure variables, you'll want to assign values to them and use them in your sketch. The next section covers how to do just that.

Using Data Structures

You can assign values to the individual data elements in a data structure in a couple of different ways. One way is to set the initial values of the data elements when you declare the structure variable:

```
struct sensorinfo morningTemps = {"01/01/14", 68, 25};
```

With this format, you just define the data elements values inside an array, using braces to create the array. When you define the array, it's important that you list the data values in the same order that you defined the data elements in the `struct` declaration.

After assigning values to the data structure, you can reference the individual data elements using this format:

```
structname.dataname
```

where *`structname`* is the data structure variable that you instantiated, and *`dataname`* is the name of the specific data element defined in the structure. For example, to retrieve the outdoor temperature from the morning, you use the following:

```
morningTemps.outdoortemp
```

You can use this format to both assign data values to the data elements or to retrieve the data currently stored in the data element. For example, to retrieve the data element values stored in the data structure, you use code like this:

```
Serial.print("The indoor temperature on ");
Serial.print(morningTemps.date);
Serial.print(" is ");
Serial.println(morningTemps.indoortemp);
```

The other way to assign values to data elements in a data structure is to use separate assignment statements for each data element:

```
struct sensorinfo morningTemps;
strcpy(morningTemps.date, "01/01/14");
morningTemps.indoortemp = 68;
morningTemps.outdoortemp = 25;
```

This example uses the `strcpy` function to copy a text value into the character array, showing that you can use the data structure element anywhere you'd use a variable.

Let's run through a quick example of using simple data structures in an Arduino sketch.

TRY IT YOURSELF ▼

Using Data Structures

In this example, you create a data structure to store multiple values to simulate a sensor reading, and then display the data structure values in your sketch output. To run the example, just follow these steps:

1. Open the Arduino IDE, and then enter this code in the editor window:

```
struct sensorinfo {
   char date[9];
   int indoortemp;
   int outdoortemp;
} morningTemps;

void setup() {
   Serial.begin(9600);
   strcpy(morningTemps.date, "01/01/14");
   morningTemps.indoortemp = 72;
   morningTemps.outdoortemp = 25;
   Serial.print("Today's date is ");
   Serial.println(morningTemps.date);
   Serial.print("The morning indoor temperature is ");
   Serial.println(morningTemps.indoortemp);
   Serial.print("The morning outdoor temperature is ");
```

```
    Serial.println(morningTemps.outdoortemp);
}

void loop() {
}
```

2. Save the sketch as **sketch0901**.
3. Click the Upload icon to verify, compile, and upload the sketch to your Arduino unit.
4. Open the serial monitor window to view the output from the sketch.

The sketch0901 sketch creates a simple data structure, instantiates it in the `morningTemps` variable, and then assigns values to each element in the data structure. It then outputs the values of the data structure elements using the standard `Serial.print` and `Serial.println` functions. Figure 9.1 shows the output you should see in the serial monitor window.

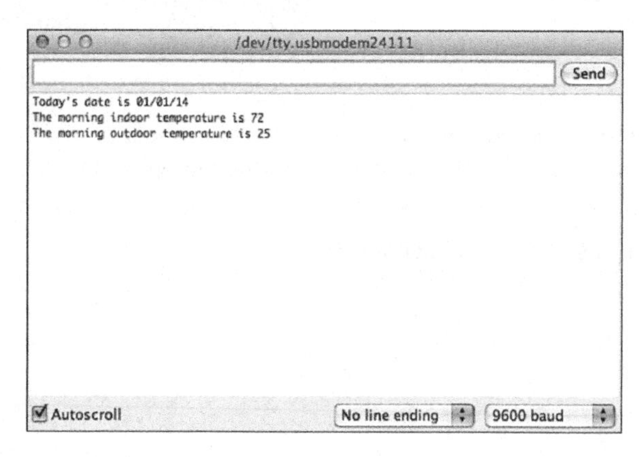

FIGURE 9.1
Output from running the sketch0901 sketch.

Manipulating Data Structures

The great thing about storing data in data structures is that now you can handle the data structure as a single object in your sketch code. For example, if you need to copy the data values from one `sensorinfo` object to another, instead of having to copy the individual data elements, you can copy the structure as a whole:

```
noonTemps = morningTemps;
```

The Arduino knows to copy the entire data structure from the morningTemps variable memory location into the noonTemps variable memory location. Here's an example that shows this in action.

TRY IT YOURSELF ▼

Copying Data Structures

In this example, you create a data structure, instantiate two occurrences of the data structure, and then copy one occurrence to the other. Here are the steps to do that:

1. Open the Arduino IDE, and then enter this code into the editor window:

```
struct sensorinfo {
   char date[9];
   int indoortemp;
   int outdoortemp;
} morningTemps, noonTemps;

void setup() {
   Serial.begin(9600);
   strcpy(morningTemps.date, "01/01/14");
   morningTemps.indoortemp = 72;
   morningTemps.outdoortemp = 25;
   noonTemps = morningTemps;
   Serial.print("For date ");
   Serial.println(noonTemps.date);
   Serial.print("The indoor temperature is ");
   Serial.println(noonTemps.indoortemp);
   Serial.print("The outdoor temperature is ");
   Serial.println(noonTemps.outdoortemp);
}

void loop() {
}
```

2. Save the file as **sketch0902**.
3. Click the Upload icon to verify, compile, and upload the sketch to your Arduino unit.
4. Open the serial monitor to run the sketch and view the output.

The data values assigned to the noonTemps data structure should be the same that you assigned to the morningTemps data structure. That shows the assignment statement copied the full data structure. Figure 9.2 shows the output you should see in the serial monitor.

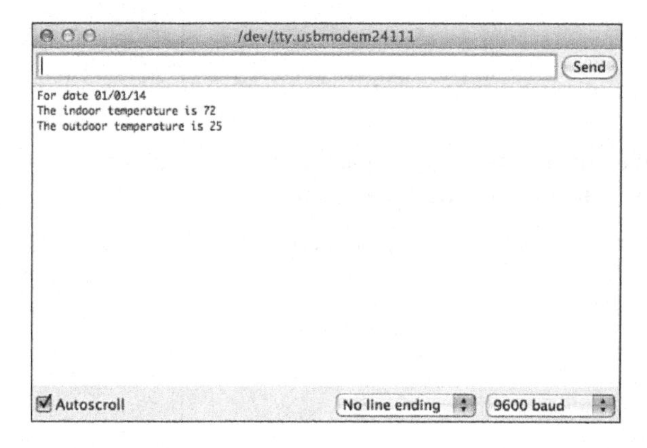

FIGURE 9.2
Output from running the sketch0902 sketch.

Arrays of Structures

You're not limited to creating a single variable data structure; you can also create arrays of a data structure:

```
struct sensorinfo temps[10];
```

This creates an array of 10 instances of the sensorinfo data structure. You can then reference each individual instance of the data structure using the standard array index format:

```
temps[0].date = "01/01/14";
temps[0].indoortemp = 72;
temps[0].outdoortemp = 25;
temps[1].date = "02/01/14";
temps[1].indoortemp = 72;
temps[1].outdoortemp = 34;
```

Now you can easily create a loop to iterate through all the instances of your data structure, just changing the index value for each instance. With this technique, you can group as many data structure instances as you need to work with in your sketch. Let's go through an example that demonstrates that.

TRY IT YOURSELF ▼

Using Data Structure Arrays

In this example, you build a sketch that stores simulated sensor data into an array of data structure variables. Here are the steps to follow:

1. Open the Arduino IDE, and then enter this code into the editor window:

```
int counter;
struct sensorinfo {
  String date;
  int outdoortemp;
  int indoortemp;
};

void setup() {
  Serial.begin(9600);
  struct sensorinfo temps[11];
  Serial.println("Loading sensor data...");
  // creating sensor data and storing in data structure
  for (counter = 1; counter <= 10; counter++) {
    temps[counter].date = "01/" + String(counter) + "/14";
    temps[counter].indoortemp = 65 + random(15);
    temps[counter].outdoortemp = 20 + random(10);
  }

  Serial.println("Retrieving sensor data...");
  for(counter = 1; counter <= 10; counter++) {
    Serial.print("For date: ");
    Serial.print(temps[counter].date);
    Serial.print(" the indoor temp was: ");
    Serial.print(temps[counter].indoortemp);
    Serial.print(" and the outdoor temp was: ");
    Serial.println(temps[counter].outdoortemp);
  }
}

void loop() {
}
```

2. Save the file as **sketch0903**.

3. Click the Upload icon to verify, compile, and upload the sketch to your Arduino unit.

4. Open the serial monitor to run the sketch and view the output.

Figure 9.3 shows the output that you should see in your serial monitor.

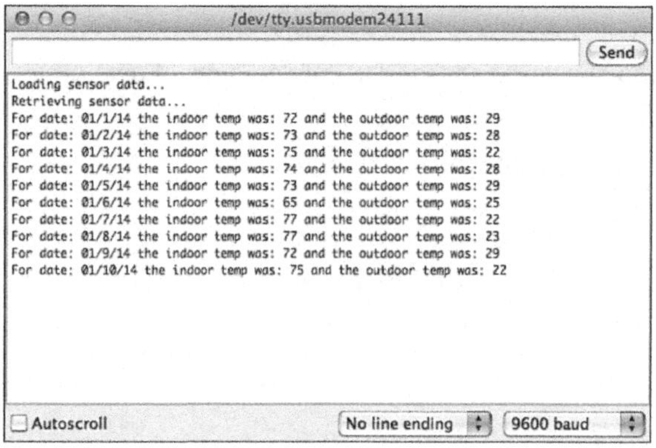

```
/dev/tty.usbmodem24111
Send
Loading sensor data...
Retrieving sensor data...
For date: 01/1/14 the indoor temp was: 72 and the outdoor temp was: 29
For date: 01/2/14 the indoor temp was: 73 and the outdoor temp was: 28
For date: 01/3/14 the indoor temp was: 75 and the outdoor temp was: 22
For date: 01/4/14 the indoor temp was: 74 and the outdoor temp was: 28
For date: 01/5/14 the indoor temp was: 73 and the outdoor temp was: 29
For date: 01/6/14 the indoor temp was: 65 and the outdoor temp was: 25
For date: 01/7/14 the indoor temp was: 77 and the outdoor temp was: 22
For date: 01/8/14 the indoor temp was: 77 and the outdoor temp was: 23
For date: 01/9/14 the indoor temp was: 72 and the outdoor temp was: 29
For date: 01/10/14 the indoor temp was: 75 and the outdoor temp was: 22
Autoscroll    No line ending    9600 baud
```

FIGURE 9.3
Output from the sketch0903 code.

Take some time to go over the code in the sketch0904 file; there are a few things to look at in this example. First, the example demonstrated how to use structure arrays, and it also used a `String` object as part of the data structure:

```
struct sensorinfo {
  String date;
  int outdoortemp;
  int indoortemp;
};
```

Then, I used a little bit of trickery to generate a date for each sample:

```
temps[counter].date = "01/" + String(counter) + "/14";
```

Because I stored the date value as a `String` object in the data structure, the value assigned to it could be a `String` structure. I took advantage of the `String` method to automatically convert the `counter` variable integer value into a string value. Making it easy to incorporate the `counter` value into the sample date.

Data structures take some time to get used to, but they're well worth the effort because they can save you quite a bit of typing if you're trying to work with lots of data in your sketches.

Working with Unions

Thanks to the C programming language, you can use one more data trick in your Arduino programs. Unions allow you to specify a variable that you can use to store a single value, but as more than one data type. This might sound confusing, and it can be somewhat odd trying to figure out just what's going on with using unions in your sketches.

First, let's take a look at how to create a union:

```
union {
   // variable list
} unionname;
```

The *variable list* used in the union differs from the variable list you use in a defining a data structure. Instead of defining multiple values, it represents the different data types the union can use to store a single value. The Arduino creates only a single memory location for the union, reserving only enough memory to store the largest data type specified.

The union variable only stores one data value, using the data type of the variable you specify. Let's take a look at an example of using a union. In this example, we'll assume we have a project with two input sensors. One sensor is a digital input that returns integer values. The other sensor is an analog input that returns floating-point values.

To define the union, you use this format:

```
union  {
   int digitalInput;
   float analogInput;
} sensorInput;
```

In this example, we don't have to worry about storing both sensor values at the same time; only one sensor will return a value at a time.

You can store the either the digital or analog input value using the `sensorInput` union variable. However, whatever one you store, you have to keep track to make sure that you retrieve the value from the union using the same data type. For example:

```
sensorInput.digitalInput = 72;
int intValue = sensorInput.digitalInput;
Serial.print("The digital input reading is ");
Serial.println(sensorInput.digitalInput);
```

If you store the value using the `digitalInput` union data element, you must retrieve it using the `digitalInput` data element. If you try to retrieve the value using the `analogInput` variable, you'll get a value, but it won't be the correct format.

The trick is that the union variable doesn't contain two separate values; it just points to one memory location that's large enough to hold the largest of the defined data types. If you store the value as an integer and try to retrieve it as a float data type, you won't get the correct result.

Let's go through an example to help reinforce this.

▼ TRY IT YOURSELF

Experimenting with Unions

In this exercise, you store a value into a union that supports two data types and then try to retrieve the value using both data types. Here are the steps to follow for the experiment:

1. Open the Arduino IDE, and then enter this code into the editor window:

```
union {
  int digitalInput;
  float analogInput;
} sensorInput;

void setup() {
  Serial.begin(9600);
  Serial.println("Saving a digital value...");
  sensorInput.digitalInput = 72;
  Serial.print("The correct digital input value is: ");
  Serial.println(sensorInput.digitalInput);
  Serial.print("The incorrect analog input value is: ");
  Serial.println(sensorInput.analogInput);

  Serial.println("Saving an analog value...");
  sensorInput.analogInput = 5.25;
  Serial.print("The incorrect digital input value is: ");
  Serial.println(sensorInput.digitalInput);
  Serial.print("The correct analog input value is: ");
  Serial.println(sensorInput.analogInput);
}

void loop() {
}
```

2. Save the file as **sketch0904**.
3. Click the Upload icon to verify, compile, and upload the sketch to your Arduino unit.
4. Open the serial monitor tool to run the sketch and view the output.

Figure 9.4 shows the output you should see from running the sketch0904 code.

When you store an integer value into the union, the `digitalInput` variable contains the correct value. However, you can still retrieve the `analogInput` variable value, but it contains an incorrect value. Likewise, when you store a float value in the union, the `analogInput` variable contains the correct value, and the `digitalInput` variable contains an incorrect value.

Unions can come in handy, but you have to be very careful when using them in your Arduino sketches.

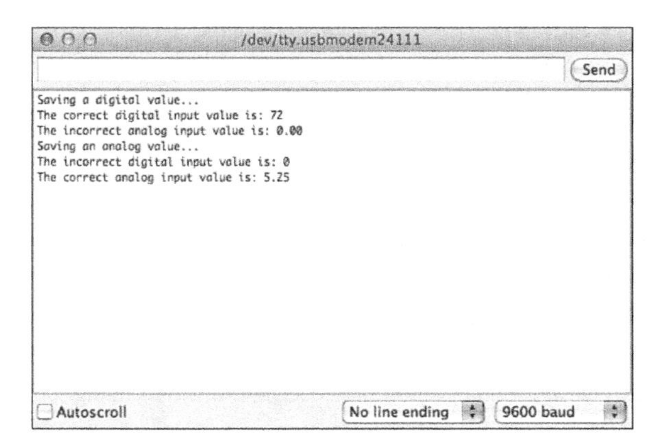

FIGURE 9.4
Output from the sketch0904 code.

Summary

This hour covered some of the more complex data handling topics in the Arduino programming language. Data structures allow us to group multiple variables together into a single object that we can handle in our sketches. You can copy structures as a single object, easily copying multiple data elements in one statement. You can also create an array of data structures, which allows you to iterate through all the instantiated values. This hour also discussed using unions in your Arduino sketches. A union allows you to store multiple data types in a single memory location, as long as you retrieve the data using the same data type as you store it.

The next hour covers how to create your own functions for your Arduino sketches. By creating functions, you can place code that you commonly use in your sketches in one place to help cut down on typing.

Workshop

Quiz

1. If you have a set of data values that you want to handle together, what's the best way to store them and access them in your sketch?
    - **A.** Use a union
    - **B.** Use an array
    - **C.** Use a data structure
    - **D.** Use multiple variables

2. If you store a value in a union using one data type and try to retrieve it using another data type, you'll get an error message. True or false?

3. How do you define an array of structures?

Answers

1. C. A data structure is designed to allow you to handle a group of variable values as a single object in your sketches. That makes it easier to manage the data that you have to work with.

2. False. The Arduino will retrieve the value stored in memory and try to interpret the value using the other data type, which won't give a meaningful answer. Be careful when using unions; it's easy to make a mistake.

3. When you instantiate a new structure variable, you place the size of the array in square brackets after the array variable:

```
struct sensorinfo temps[20];
```

Q&A

Q. Is there a limit to how many data elements I can place in a data structure?

A. Not in theory. You can store as many data elements in a structure as necessary. Just remember that they each take up space in memory, and there is a limit to how much memory is in the Arduino.

Q. Can I use a union to convert a value from one data type to another?

A. No. The Arduino doesn't convert the data type stored in the memory location to the type you request. It assumes that the data stored in the memory location is in the proper format as the request and interprets it as such. If the data type is not correct, the resulting value will be incorrect.

HOUR 10
Creating Functions

What You'll Learn in This Hour:

- How to create your own functions
- Retrieving data from functions
- Passing data to functions
- Using functions in your Arduino sketches

While writing Arduino sketches, you'll often find yourself using the same code in multiple locations. With just a small code snippet, it's usually not that big of a deal. However, rewriting large chunks of code multiple times in your Arduino sketches can get tiring. Fortunately, Arduino provides a way to help you out by supporting user-defined functions. You can encapsulate your C code into a function and then use it as many times as you want, anywhere in your sketch. This hour discusses the process of creating your own Arduino functions, and demonstrates how to use them in other Arduino sketch applications.

Basic Function Use

As you start writing more complex Arduino sketches, you'll find yourself reusing parts of code that perform specific tasks. Sometimes it's something simple, such as retrieving a value from a sensor. Other times, it's a complicated calculation used multiple times in your sketch as part of a larger process.

In each of these situations, it can get tiresome writing the same blocks of code over and over again in your sketch. It would be nice to just write the block of code once and then be able to refer to that block of code anywhere in your sketch without having to rewrite it.

Arduino provides a feature that enables you to do just that. Functions are blocks of code that you assign a name to and then reuse anywhere in your code. Anytime you need to use that block of code in your sketch, you simply use the function name you assigned to it (referred to as *calling the function*). This section describes how to create and use functions in your Arduino sketches.

Defining the Function

The format that you use to create a function in Arduino doesn't use a keyword as some other programming languages. Instead, you declare the data type that the function returns, along with the function name with parentheses:

```
datatype funcname() {
    // code statements
}
```

The *funcname* defines the name of the function that you use to reference it in your Arduino sketch. Function names follow the same rules as variable names, although often developers like to start functions with capital letters to help differentiate them from variables. The *datatype* defines the data type of the value the function returns (more on that later).

You must surround the code statements that you include in the function with opening and closing braces. This defines the boundaries of the code that the Arduino runs for the function.

WATCH OUT

Placing the Function Definition

Make sure that you define the function outside of the `setup` and `loop` functions in your Arduino sketch code. If you define the function inside another function, the inner function becomes a local function, and you can't use it outside of the outer function.

To create a function that doesn't return any data values to the calling program, you use the `void` data type for the function definition:

```
void MyFunction() {
   Serial.println("This is my first function");
}
```

After you define the function in your Arduino sketch, you're ready to use it.

Using the Function

To use a function that you defined in your sketch, just reference it by the function name you assigned, followed by parentheses:

```
void setup() {
   Serial.begin(9600);
   MyFunction();
   Serial.println("Now we're back to the main program");
}
```

When the Arduino runs the sketch and gets to the function line, it jumps to the function code, runs it, and then returns back to the next line in main program to continue processing. Let's take a look at an example that demonstrates how this works.

TRY IT YOURSELF ▼

Using Functions

For this example, you create a function in your Arduino sketch, and then use it in the `setup` function code for the sketch. To create the example, follow these steps:

1. Open the Arduino IDE, and enter this code into the editor window:

```
void setup() {
   Serial.begin(9600);
   Serial.println("This is the start of the main program");
   MyFunction();
   Serial.println("Now we're back to the main program");
   MyFunction();
   Serial.println("This is the end of the test");
}

void loop() {
}

void MyFunction() {
   Serial.println("This is the code from the function");
}
```

2. Save the sketch code as **sketch1001**.
3. Click the Upload icon to verify, compile, and upload the sketch to your Arduino unit.
4. Start the serial monitor to run and view the output from your sketch.

The example creates a simple function called `MyFunction` that outputs a line of text to the serial monitor. The setup function displays some text so that you know that it's starting, calls the function, and then displays some more text out so that you know the sketch returned back to the main program. If all goes well, you should see the output shown in Figure 10.1.

Congratulations, you've just created a function. If you have to repeat a long message in your sketch, such as help information, you can create a function to do that, and just call the sketch every time you need the output.

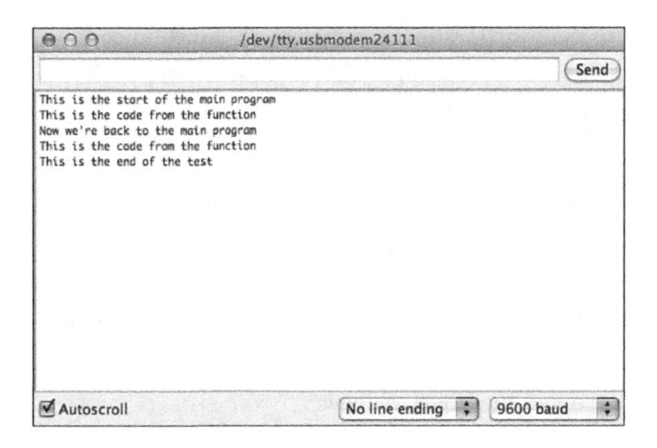

FIGURE 10.1
The output from the sketch1001 code.

Returning a Value

You can do a lot more in functions than just output text. Functions can process any statements that the main program can. This allows you to create functions to process repetitive data calculations, and return a result from the calculations.

To return a value from the function back to the main program, you end the function with a `return` statement:

```
return value;
```

The *`value`* can be either a constant numeric or string value, or a variable that contains a numeric or string value. However, in either case, the data type of the returned value must match the data type that you use to define the function:

```
int MyFunction2() {
   return 10 * 20;
}
```

If you use a variable to hold the returned value, you must declare the variable data type as the same data type used to declare the function:

```
float MyFunction3() {
   float result;
   result = 10.0 / 2.0;
   return result;
}
```

To retrieve the value returned from a function, you must assign the output of the function to a variable in an assignment statement:

```
value = MyFunction3();
```

Whatever data value the function returns is assigned to the `value` variable in your sketch.

Let's take a look at an example of using a return value from a function in an Arduino program.

TRY IT YOURSELF 

Returning a Value from a Function

In this example, you create a function to perform a mathematical operation, and then return the value back to the main program. Here are the steps to do that:

1. Open the Arduino IDE and enter this code into the editor window:

```
void setup() {
  int value;
  Serial.begin(9600);
  value = calc();
  Serial.print("The result from the function is ");
  Serial.println(value);
}

void loop() {
}

int calc() {
  int result;
  result = 2 * 2 + 10;
  return result;
}
```

2. Save the sketch as **sketch1002**.
3. Click the Upload icon to verify, compile, and upload the sketch to your Arduino unit.
4. Open the serial monitor tool to view the output from the sketch.

This example sketch creates a simple function called `calc`, which performs a mathematical calculation and then returns the value back to the program as an integer data type. The main sketch code in the `setup` function calls the `calc` function and assigns the returned value to the value variable. When you run this sketch, you should get the output shown in Figure 10.2.

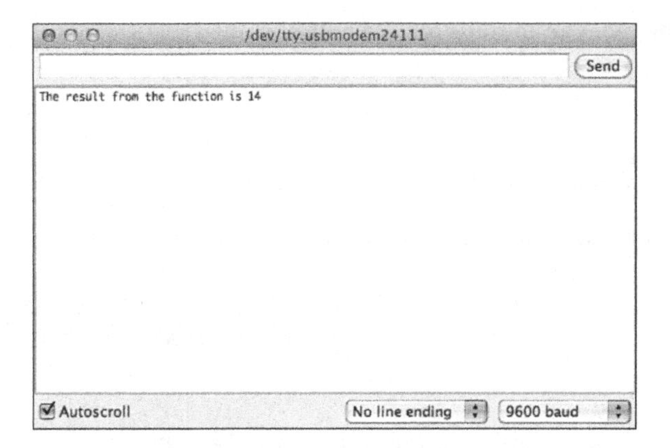

FIGURE 10.2
The output from running the sketch1002 code.

Returning a value from a function is handy, but you'll usually want to be able to perform some type of operation in the function to allow values to also be passed into the function. The next section shows how to do that.

Passing Values to Functions

There's just one more piece to making the most use out of functions in your Arduino sketches. Besides returning a value from a function, you'll most likely want to be able to pass one or more values into the function for processing.

In the main program code, you specify the values passed to a function in what are called arguments, specified inside the function parentheses:

```
value = area(10, 20);
```

The `10` and `20` value arguments are listed separated by a comma. If you just have one argument to pass, you don't use the comma.

To retrieve the arguments passed to a function, the function definition must declare what are called parameters. You do that in the main function declaration line:

```
int area(int width, int height) {
```

The parameter definitions define both the data type of the parameter, and a variable name. The argument data values are assigned to the parameter variable names in the same order that they're listed in the function declaration. In this example, therefore, the `width` variable is assigned the value `10` and the `height` variable is assigned the value `20` inside the function code.

You can use the parameter variables inside the function just as you would any other variable:

```
int area(int width, int height) {
   int result = width * height;
   return result;
}
```

After you define the function with parameters, you can use the function as often as you need in your sketch code, passing different argument values to the function to retrieve different results. Let's take a look at an example of doing just that.

TRY IT YOURSELF ▼

Passing Arguments to Functions

In this example, you create a function that you can use multiple times in your Arduino sketch, passing different argument values and retrieving different results. To try out this example, follow these steps:

1. Open the Arduino IDE and enter this code into the editor window:

```
void setup() {
  int returnValue;
  Serial.begin(9600);

  Serial.print("The area of a 10 x 20 size room is ");
  returnValue = area(10, 20);
  Serial.println(returnValue);

  Serial.print("The area of a 25 x 15 size room is ");
  returnValue = area(25, 15);
  Serial.println(returnValue);
}

void loop() {
}

int area(int width, int height) {
  int result = width * height;
  return result;
}
```

2. Save the sketch as **sketch1003**.
3. Click the Upload icon to verify, compile, and upload the sketch to your Arduino unit.
4. Open the serial monitor to view the output from the sketch.

The `area` function retrieves the two parameters passed to the function, multiplies them, and then assigns the result to the `result` variable. The `area` function then uses the `return` statement to return the result back to the calling code. The `setup` function code uses the `area` function twice, with two different sets of values. Figure 10.3 shows the result you should see in your serial monitor after running the sketch.

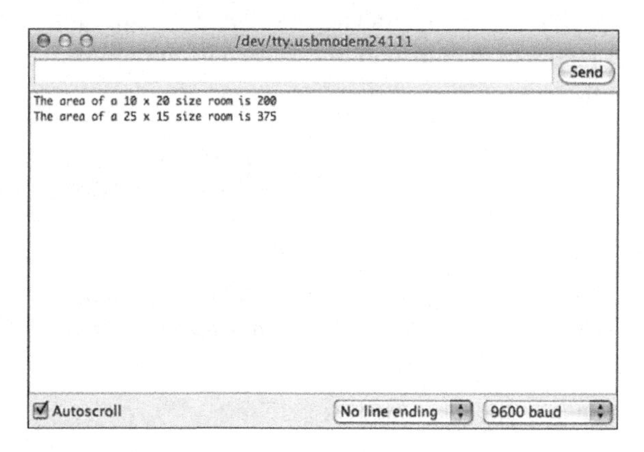

FIGURE 10.3
The serial monitor output from running the sketch1003 code.

Now you have all the tools you need to create and run your own functions in your Arduino sketches. You'll want to know about just a couple of other items to help make your life a little easier when working with functions. One of those is how to declare variables in your sketch, which the next section discusses.

Handling Variables in Functions

One thing that causes problems for beginning sketch writers is the scope of a variable. The scope is where the variable can be referenced within the sketch. Variables defined in functions can have a different scope than regular variables. That is, they can be hidden from the rest of the sketch.

Functions use two types of variables:

- Global variables
- Local variables

The following sections describe how to use both types of variables in your functions.

Defining Global Variables

Global variables are variables that are valid anywhere within the sketch code. If you define a global variable in the main section of your sketch, outside of the functions, you can retrieve its value inside any function in the sketch.

Let's take a look at a simple example of defining a global variable:

```
const float pi = 3.14;

void setup() {
  Serial.begin(9600);
  float area;

  Serial.print("The area of a circle with radius of 5 is ");
  area = circleArea(5);
  Serial.println(area);
  Serial.print("The value of pi used for this calculation was ");
  Serial.println(pi);
}

void loop() {
}

float circleArea(int radius) {
  float result = pi * radius * radius;
  return result;
}
```

Notice in the sketch, the `pi` variable is defined before any of the functions are defined. That makes it a global variable, and you can use it anywhere in the sketch code, including the `setup` and `loop` functions.

This can be a dangerous practice, however, especially if you intend to use your function in different sketches. The function code assumes the global variable exists, and will fail if it doesn't.

Another danger of using global variables is if you use a global variable inside a function for a different purpose than what you use it outside of the function. If the function changes the global variable value and other functions don't expect that, it might cause problems in your sketch. Here's an example of this problem:

```
int temp, value;

void setup() {
   int result;
   Serial.begin(9600);
   temp = 4;
   value = 6;
```

```
    result = func1();
    Serial.print("The result is ");
    Serial.println(result);
    Serial.print("The value of temp is ");
    Serial.println(temp);
}

void loop() {
}

int func1() {
    temp = value + 5;
    return temp * 2;
}
```

In this example, the `temp` and `value` variables are defined as global variables, and can be used in any function in the sketch. The `temp` variable was used in both the `setup` and `func1` function, but for different purposes.

In the `setup` function, the `temp` variable is set to a static value of `4`; however, it's also used in the `func1` function as a temporary variable, changing the value for a calculation. When the sketch cod returns back to the `setup` function, the value stored in the `temp` variable is now different. If you run this example in the serial monitor, you'll get the following output:

```
The result is 22
The value of temp is 11
```

The change made to the `temp` variable in the `func1` function affected the value in the entire sketch, not just in the `func1` function. If you were expecting the `temp` variable to still be set to `4`, you'd run into a problem with your code.

If you do need to use temporary variables inside a function, there's a better way to do that than using global variables, as shown in the next section.

Declaring Local Variables

Instead of using global variables in functions, which run the risk of conflicting with other functions, you can use a local variable. You declare a local variable inside the function code itself, separate from the rest of the sketch code:

```
int func1() {
    int calc;
    calc = value + 5;
    return calc * 2;
}
```

Once you declare the calc variable as a local variable inside the function, you can use it only inside the function where you declare it. If you try using it in another function, you'll get an error:

```
sketch10test.ino: In function 'void setup()':
sketch10test:12: error: 'calc' was not declared in this scope
```

Another interesting feature of local variables is that you can override a global variable with a local variable:

```
int temp, value;

void setup() {
   int result;
   Serial.begin(9600);
   temp = 4;
   value = 6;
   result = func1();
   Serial.print("The result is ");
   Serial.println(result);
   Serial.print("The value of temp is ");
   Serial.println(temp);
}

void loop() {
}

int func1() {
   int temp;
   temp = value + 5;
   return temp * 2;
}
```

In this example, the temp variable is declared a second time in the func1 function as a local variable. That makes it separate from the temp variable that is declared as a global variable. Now if you run this sketch, you'll get the following output:

```
The result is 22
The value of temp is 4
```

So, the original value assigned to the temp global variable remained intact after the function.

WATCH OUT

Overriding Global Variables

You can override a global variable with a local variable, but it's not a good practice. When you use the same variable name for both a global and local variable, it can make trying to follow the sketch extremely difficult.

Calling Functions Recursively

One feature that local function variables provide is self-containment. A self-contained function doesn't use any resources outside of the function other than whatever variables the sketch passes to it as arguments.

This feature enables the function to be called recursively, which means that the function calls itself to reach an answer. Usually, a recursive function has a base value that it eventually iterates down to. Many advanced mathematical algorithms use recursion to reduce a complex equation down one level repeatedly until they get to the level defined by the base value.

Let's walk through an example that uses a recursive function to calculate the factorial of a number.

▼ TRY IT YOURSELF

Creating a Factorial Function

A factorial of a number is the value of the preceding numbers multiplied with the number. So, to find the factorial of 5, you perform the following equation:

```
5! = 1 * 2 * 3 * 4 * 5 = 120
```

Using recursion, the equation is reduced down to the following format:

```
x! = x * (x - 1)!
```

Or in English: The factorial of `x` is equal to `x` times the factorial of `x - 1`. By definition, the factorial of 1 is 1, so that's the base value the recursion will stop at. This can be expressed in a simple recursive function:

```
int factorial(int x) {
   int result;
   if (x == 1)
      return 1;
   else {
      result = x * factorial(x - 1);
```

```
      return result;
   }
}
```

Now just follow these steps to use that function in a sketch:

1. Open the Arduino IDE and enter this code into the editor window:

```
void setup() {
  Serial.begin(9600);
  int fact, value;

  fact = factorial(5);
  Serial.print("The factorial of 5 is ");
  Serial.println(fact);

  fact = factorial(7);
  Serial.print("The factorial of 7 is ");
  Serial.println(fact);
}

void loop() {
}

int factorial(int x) {
  int result;
  if (x == 1)
     return 1;
  else   {
    result = x * factorial(x - 1);
    return result;
  }
}
```

2. Save the sketch code as **sketch1004**.
3. Click the Upload icon to verify, compile, and upload the sketch into your Arduino unit.
4. Open the serial monitor to view the output of the sketch.

Figure 10.4 shows the output that you should see in the serial monitor.

The `factorial` recursive function iterates through the function until it gets to the base value for the factorial and then returns the final result back to the main calling program. Because the function is completely self-contained, you can use the `factorial` function in any sketch that requires a factorial calculation.

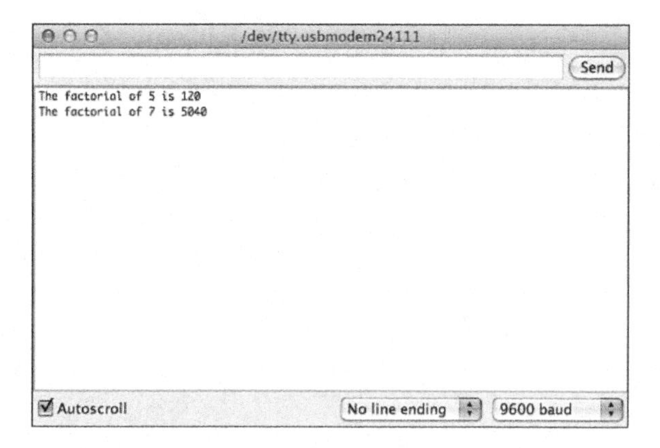

FIGURE 10.4
Output from running the sketch1004 code.

Summary

This hour showed how to create and use functions in your Arduino sketches. You can define functions to help reduce the amount of coding you have to repeat in your sketch. You can define a function to return value back to the main program, or it can just return back with no return value. You can also pass one or more argument values to a function to be used inside the function code.

The next hour covers one of the more complex features of C programming that you can also use in your Arduino programs. C pointers can get confusing, but once you get the hang of them, they can come in handy in your sketches.

Workshop

Quiz

1. Which format is correct for defining a function in an Arduino sketch?

   A. `def MyFunction()`

   B. `function myFunction()`

   C. `MyFunction()`

   D. `void MyFunction()`

2. You can use a variable name as both a global variable and a local variable and the values will be stored in the same location. True or false?

3. What feature of functions can you use to reduce a complex formula down to multiple iterations of decreasing complexity to return an answer?

Answers

1. D. In the Arduino programming language, you must define the data type the function returns when you declare the function. The `void` data type indicates the function doesn't return a value.

2. False. The Arduino will reserve a separate memory location for the local variable from the global variable, and will treat the two variables as separate objects.

3. Recursion allows a function to call itself with a subset of the equation, reducing the problem down to a base result. Once the recursion reaches the base result, the function returns the answers back through the iterations to return the answer.

Q&A

Q. Can a function call other functions inside the function code?

A. Yes, you can call other functions from inside a function.

Q. Is there a limit to how many times a recursive function can call itself?

A. No, in the next hour, ion can continue on as many times as necessary. Of course, this can lead to an endless loop of recursion, so be careful to make sure that your recursive functions have a base value to end on.

Q. Can you write a library to store common functions and then reference that library in any sketch?

A. Yes, this is called a function library. You'll learn how to create and use a function library later on in Hour 13, "Using Libraries."

HOUR 11
Pointing to Data

What You'll Learn in This Hour:

- What C pointers are
- How to define a pointer in your sketch
- How to use pointers
- When to use pointers in different sketch environments

By far, one of the most complicated topics in the C programming language is data pointers. Although it's quite possible that you'll never have to use them in your sketches, data pointers can allow you to do some pretty fancy things in your Arduino sketches, sometimes reducing the amount of code you have to write. This hour delves into the world of C data pointers and shows how you can use them in your Arduino sketches.

What Is a Pointer?

Hour 5, "Learning the Basics of C," showed how to use variables in your Arduino sketches for storing data in memory. When you define a variable in your sketch, the Arduino system selects a location in the static random-access memory (SRAM) to store the data value that the variable represents.

Most of the time, we don't care where the Arduino stores the data values in memory, just as long as it can keep track of the data for us. However, it can come in handy knowing where in memory the Arduino stored the variable so that you can directly access the data yourself, sometimes using less code.

To do that, the C programming language provides pointers. Pointers allow us to take a peek under the hood and see just where the Arduino stores the data in our sketches, and then use that information to directly access your data.

You use two operators to handle pointers in Arduino sketches:

- & (called the reference operator)
- * (called the dereference operator)

The reference operator returns the memory location where the Arduino stores a variable value. Just place the reference operator in front of any variable to return the memory address where the Arduino stores the variable's value:

```
ptr = &test;
```

The `ptr` variable now contains the memory address location where the Arduino placed the test variable. The problem is that the Arduino returns the memory address as a special value. This is where the dereference operator comes into play.

You use the dereference operator to identify a pointer variable used to store memory locations:

```
int test = 100;
int *ptr;
ptr = &test;
```

The `ptr` pointer variable now contains a value that represents the memory location where the `test` variable value is stored. You can display that value using the `Serial.print` function just as you would any other variable value. Memory locations are often displayed as hexadecimal values. To do that, just add the `HEX` parameter in the `Serial.print` function:

```
Serial.print(ptr, HEX);
```

Notice that I declared the `ptr` variable using the same data type as the variable that it points to in memory. There's a reason for that, as discussed a little later on in this hour. However, because of that, if you need to retrieve the memory locations of different variables of different data types, you should use the appropriate data type for each associated pointer:

```
float test2 = 3.14159;
char test3[40] = "This is a test character array string";
float ptr2 = &test2;
char ptr3 = &test3;
```

The `ptr2` and `ptr3` variables each contain the memory address location for the individual data variable they point to. Let's practice handling pointers and displaying them in the serial monitor.

TRY IT YOURSELF ▼

Finding Your Data in Memory

You can use the reference operator in your Arduino sketch to find the memory location the Arduino uses to store a variable value. Here's an experiment you can run to test this:

1. Open the Arduino IDE and enter the following code into the editor window:

```
int test1 = 10;
float test2 = 3.14159;
char test3[] = "This is a character array";
int *ptr1;
float *ptr2;
char *ptr3;
char buffer[50];

void setup() {
   Serial.begin(9600);
   ptr1 = &test1;
   sprintf(buffer, "The integer value is stored at %x", ptr1);
   Serial.println(buffer);
   ptr2 = &test2;
   sprintf(buffer, "The float value is stored at %x", ptr2);
   Serial.println(buffer);
   ptr3 = test3;
   sprintf(buffer, "The character array value is stored at %x", ptr3);
   Serial.println(buffer);
}

void loop() {
```

2. Save the sketch as **sketch1101**.
3. Click the Upload icon to verify, compile, and upload the sketch to your Arduino unit.
4. Open the serial monitor to run the sketch and view the output.

When you run sketch1101 and look at the output in the serial monitor, you'll see the actual memory locations where your Arduino stores each variable value. Figure 11.1 shows the output from my Arduino unit. (Your memory locations may vary from mine.)

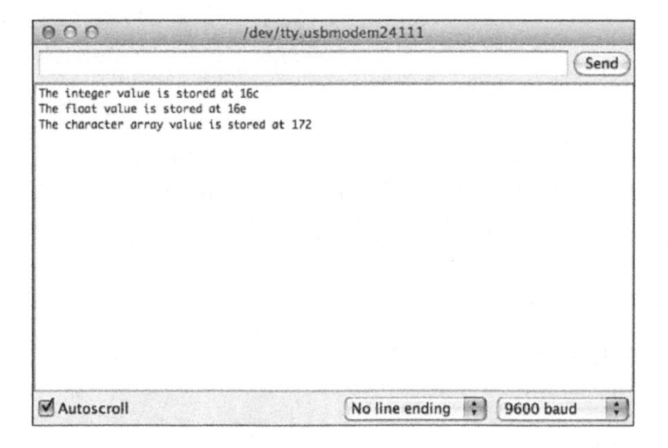

FIGURE 11.1
The output from sketch1101 showing the memory locations.

BY THE WAY

Printing Pointers

Unfortunately, the `Serial.println` function in the Arduino code doesn't support printing pointer values. To get around that, the sketch1101 code stores the output in a character array buffer using the `sprintf` C function. The `%x` code tells the Arduino to output the value in hexadecimal notation.

You may see that the Arduino placed the variables in sequence within the memory area. (There should be enough separation between each location to hold the number of bytes required for each data type: 2 bytes for the integer value, and 4 bytes for the floating-point value.)

Working with Pointers

Now that you've seen the basics of how the reference and dereference operators work, let's take a look at how to use them in sketches. This section walks through the details of how to use pointers in sketches.

Retrieving Data Using a Pointer

Once you have a memory location stored in a pointer variable, you can use that information to extract the value stored at that location. You just use the dereference operator to do that:

```
int test1 = 10;
int *ptr = &test1;
int test2 = *ptr;
```

The `ptr` pointer variable contains the memory address for the `test1` variable. To retrieve the value stored at that memory location, just use the dereference operator on the `ptr` variable. The `test2` assignment statement assigns the value stored in the memory address pointed to by the `ptr` pointer variable to the `test2` variable. The `test2` variable will now have to value `10`.

Storing Data Using a Pointer

Likewise, you can assign a value to a memory location using the dereference operator:

```
int test1 = 100;
int *ptr = &test1;
*ptr = 200;
```

The `ptr` pointer variable points to the memory location used to store the `test1` variable value. When you assign a value to the dereferenced memory location, this places the value into the memory location. After this code runs, the `test1` variable will contain the value of `200`.

Using Special Types of Pointers

You can run into a couple of special situations when using pointers. This section walks through two situations that might seem odd but that can come in handy when using pointers in your Arduino sketches.

Null Pointers

When you declare a pointer but don't initialize it to a specific memory location, the pointer may or may not have an initial value assigned to it by the Arduino. That can cause issues in your sketch if you're not careful. One way to avoid that is to declare a null pointer. A null pointer is set to a value of `0`, which doesn't represent a valid location in memory. You do that using the special `NULL` label in C:

```
int *ptr1 = NULL;
```

The `NULL` value assigns a value of `0` to the pointer `ptr1` as the initial value.

One benefit of initializing an empty pointer to the `NULL` value is that it allows you to easily check to see whether a pointer is in use or not in your sketch code by using an `if-then` condition check:

```
if (ptr1) {
   // pointer in use
} else {
   // pointer is NULL
}
```

If the pointer has a NULL value assigned to it, it returns a false value from the if-then condition check.

Void Pointers

Another pointer trick that can come in handy is to use the void data type when declaring a pointer:

```
void *ptr1 = NULL;
```

This is called a void pointer. Because there isn't a data type associated with the pointer, you can use the same pointer variable for storing the memory address of different data type variables:

```
int test1 = 10;
float test2 = 3.14159;
char[] test3 = "This is a character array string";
void *ptr = NULL;
ptr = &test1;
Serial.print(ptr, HEX);
ptr = &test2;
Serial.print(ptr, HEX);
ptr = &test3;
Serial.print(ptr, HEX);
```

You can use the void data type to represent a pointer if you're not sure of the specific data type of the data you need to point to. However, that does limit what you can do with the pointer somewhat. When the Arduino knows the data type that the pointer points to, you can use that information to perform some fancy data manipulation. The next section shows how to do that.

Pointer Arithmetic

One popular use of pointers in the C programming language is manipulating values stored in an array variable. If you remember from Hour 7, "Programming Loops," array variables store multiple data values sequentially in memory, with each individual data value referenced using an index value. For example:

```
int test[5] = {10, 20, 30, 40, 50};
Serial.println(test[0]);
```

You reference a specific data value stored in the array variable using the index value between the square brackets. The index values start at 0, and go to the end of the array data values.

Because pointer variables point to locations in memory, you can use a pointer to point to the location of the start of the array variable in memory:

```
int *ptr = &test[0];
```

This line uses the dereference operator to return the memory location where the Arduino stores the first data value in the test array.

Now here's where the data type for the pointer comes into play. If you increment the `ptr` value by one, the Arduino actually makes it point to the next data element in memory for that data type. Thus, if you do this:

```
ptr++;
```

The `ptr` variable will now point to the memory location where the `test[1]` array variable value is stored.

Now here's the really cool part. The pointer arithmetic works for any data type you use. If you create an array of float values and create a pointer to it

```
float test2[] = {1.23, 4.56, 7.89};
float *ptr2 = &test2[0];
```

the pointer arithmetic increments the pointer to the next floating-point number in the array, based on the size of the `float` data type. So, when you use the statement

```
ptr2++;
```

the Arduino skips to the next `float` value stored in the `test2` array.

This feature allows you to iterate through an array of any data type without using an array index value. Let's go through an example of doing that in an Arduino sketch.

TRY IT YOURSELF ▼

Using Pointers with Arrays

This example uses a pointer to create an array of integer values, and then uses the pointer to iterate through the stored values. Here are the steps to run the example:

1. Open the Arduino IDE, and enter this code into the editor window:

```
void setup() {
   Serial.begin(9600);
   int i;
   int test[10];
   int *ptr = test;
   *ptr = 1;
   ptr++;
   *ptr = 2;
   ptr = ptr + 1;
   *ptr = 3;
```

```
        ptr++;
        *ptr = 4;
        for(i = 0; i < 4; i++) {
            Serial.print("The value in test[");
            Serial.print(i);
            Serial.print("] is ");
            Serial.println(test[i]);
        }
    }
    void loop() {
    }
```

2. Save the code as **sketch1102**.
3. Click the Upload icon to verify, compile, and upload the sketch to your Arduino unit.
4. Open the serial monitor to run the sketch and view the output.

Figure 11.2 shows the output that you should get from running the sketch.

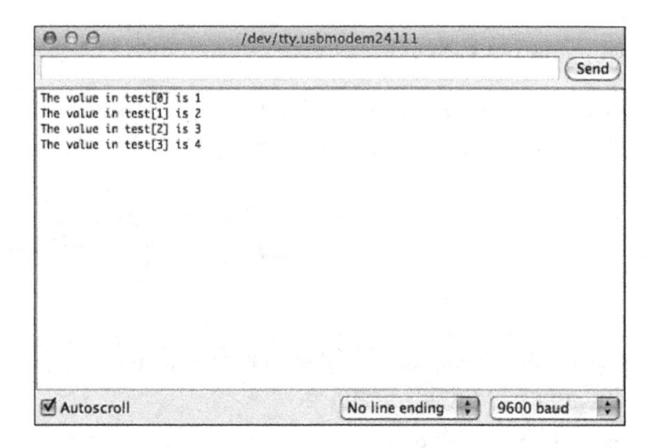

FIGURE 11.2
Output from running the sketch1102 code.

We were able to store values in the array by directly accessing the individual data elements using the pointers. The sketch code uses one line that you may not be familiar with:

```
int *ptr = test;
```

We defined the test variable as an array, but in this statement we don't use the index value for the array. This format is a shortcut to using the reference operator with the array variable:

```
int *ptr = &test[0];
```

By referencing just the array variable name without an index value, the Arduino knows you're referencing the memory location instead of a value in the array.

Strings and Pointers

Another popular use of pointers is working with character array string values. Just like a numeric array, the pointer can reference the starting memory location of the character array. You can use this for a couple of tricks.

Using Pointers for String Manipulation

First, you can use a pointer to reference the individual values in the character array to perform operations on an individual character instead of the entire string. To demonstrate this, let's go through an example of using pointers to copy a character array string one character at a time.

TRY IT YOURSELF 

Using Pointers to Copy a String

You can use pointers to reference individual characters within a string value in your sketches. This example defines two strings, and then uses two pointers to copy the characters from one array into the other. Follow these steps to see the example in action:

1. Open the Arduino IDE, and enter this code into the editor window:

```
void setup() {
   Serial.begin(9600);
   char string1[] = "This is a test string";
   char string2[22];
   char *ptr1 = string1;
   char *ptr2 = string2;
   while(*ptr1 != '\0') {
      *ptr2 = *ptr1;
      ptr2++;
      ptr1++;
   }
   Serial.print("The result is: ");
   Serial.println(string2);
}

void loop() {
}
```

2. Save the sketch as **sketch1103**.
3. Click the Upload icon to verify, compile, and upload the sketch to your Arduino unit.
4. Open the serial monitor tool to run the sketch and view the output.

The sketch uses the `ptr1` and `ptr2` pointers to directly access the individual characters within both character arrays. In the `while` loop, the code copies the character from one string into the other, and then increments both pointers. Because we know the string value terminates with a `NULL` character, we just look for that character in the `while` condition. Figure 11.3 shows the output you should see in the serial monitor window.

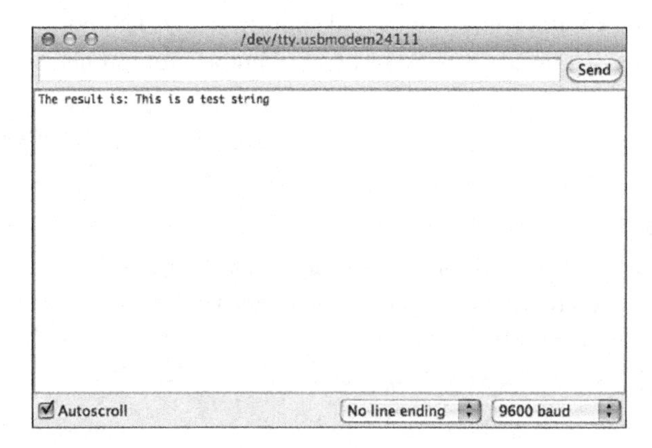

FIGURE 11.3
Output from copying a character array using a pointer.

Using Pointers to Reference Strings

Another popular use for pointers with strings is as a shortcut when declaring a string value:

```
char *string3 = "This is another test";
```

When you make this assignment, the Arduino knows to place the string into a memory location as a character array, terminate it with a `NULL` character, and then assign the starting memory location to the `string3` pointer variable. You can then refer to the string variable in your sketch to display the string:

```
Serial.print(string3);
```

What's really cool is that you can then create an array of string values to use in your sketch:

```
char *messages[3] = {"Sorry, that's too low", "Sorry, that's too high",
"That's the correct answer");
```

Then you can just use the index value to display the appropriate message that you need:

```
Serial.print(messages[2]);
```

This makes working with strings easier than trying to keep track of a lot of separate character array variables.

Combining Pointers and Structures

One of the most complicated and often misunderstood uses of pointers is applying them to data structures. Hour 9, "Implementing Data Structures," showed how to create a data structure to hold multiple data elements.

Just as with the standard data types, you can create a pointer to point to the start of the memory location where a data structure is stored. The tricky part comes in how to reference the individual data elements inside the data structure.

If you remember from Hour 9, normally you reference a structure data element using the dot notation. For example:

```
struct sensorInfo (
   char date[9];
   int indoortemp;
   int outdoortemp;
};
struct sensorInfo morningTemps, noonTemps, eveningtemps;
strcpy(morningTemps.date, "01/01/14");
morningTemps.indoortemp = 72;
morningTemps.outdoortemp = 23;
```

The `morningTemp` variable represents an instance in memory of the `sensorInfo` data structure, and the `morningTemp.outdoortemp` variable represents an individual data value stored in the structure.

You can create a pointer to the `sensorInfo` data structure using the dereference operator:

```
struct sensorInfo *testTemps;
```

Now the `testTemps` variable points to a location in memory to store a `sensorInfo` data structure. To access the individual data elements inside the structure, you could use this format:

```
*(testTemps).indoortemp
```

However, the C developers decided that was a bit awkward, so they provided a second way to access individual data elements from a data structure pointer:

```
testTemps->indoortemp
```

The `->` operator indicates the `testTemps` variable is a pointer, and that we're trying to access the data value of the `indoorTemp` element within the structure. Let's go through a simple example to demonstrate using this in a real Arduino sketch.

▼ TRY IT YOURSELF

Using Pointers with Structures

This example demonstrates how to both store and retrieve data from a data structure using a pointer. Just follow these steps to work out the example:

1. Open the Arduino IDE, and enter this code into the editor window:

```
void setup() {
    Serial.begin(9600);
    struct sensorInfo {
        char  date[9];
        int indoortemp;
        int outdoortemp;
    } morningTemps, *ptr;

    strcpy(morningTemps.date, "01/01/14");
    morningTemps.indoortemp = 75;
    morningTemps.outdoortemp = 25;

    ptr = &morningTemps;
    Serial.print("For ");
    Serial.println(ptr->date);
    Serial.print("The indoor temp was: ");
    Serial.println(ptr->indoortemp);
    Serial.print("The outdoor temp was: ");
    Serial.println(ptr->outdoortemp);

    strcpy(ptr->date, "01/02/13");
    ptr->indoortemp = 70;
    ptr->outdoortemp = 20;
    Serial.print("For ");
    Serial.println(ptr->date);
    Serial.print("The indoor temp was: ");
    Serial.println(ptr->indoortemp);
```

```
        Serial.print("The outdoor temp was: ");
        Serial.println(ptr->outdoortemp);
    }

    void loop() {
    }
```

2. Save the sketch code as **sketch1104**.
3. Click the Upload icon to verify, compile, and upload the sketch to your Arduino unit.
4. Open the serial monitor to run the sketch and view the output.

When the sketch code runs, you should see the output shown in Figure 11.4.

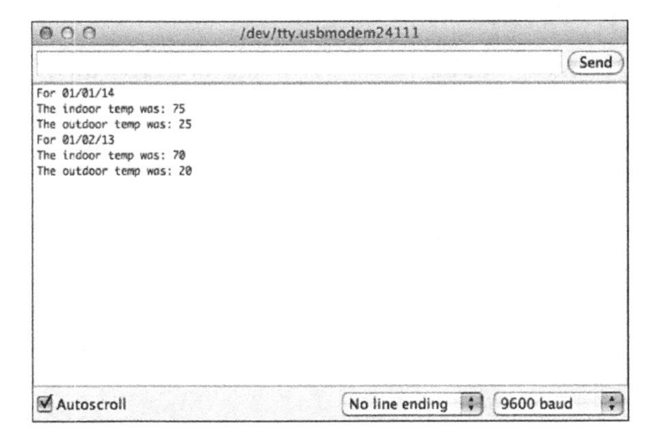

FIGURE 11.4
Using pointers with data structures.

When you define the data structure, the `morningTemps` variable points to the data structure instance in memory, and the `ptr` variable stores a memory location that points to the data structure.

The code retrieves the location of the `morningTemps` variable from memory, and assigns it to the `ptr` pointer variable:

```
ptr = &morningTemps;
```

The code then uses the `->` operator to both retrieve the data values stored in the data structure

```
Serial.println(ptr->date);
```

and to store new values into the data structure:

```
ptr->indoortemp = 70;
```

WARNING

Initializing Structures Using Pointers

You can define the values of a data structure variable in the assignment statement:

```
struct sensorInfo morningTemps = {"01/01/14", 75, 25"};
```

However, you can't do that with a pointer variable. The structure pointer variable must point to an existing instance of the data structure in memory.

Using Pointers with Functions

One last pointer topic to cover is how to use pointers with functions. One of the downsides to functions is that the return statement can only return one value back to the calling sketch code. In your sketch writing, you may run into a situation where you need to return more than one value.

One way to solve that problem is to use global variables. However, you saw in Hour 10, "Creating Functions," that makes the function code reliant on resources external to the function.

Pointers help us solve that problem. When you define a function, instead of passing arguments as values to the function, you can define the function to accept pointers instead. The pointers point to the location in memory where the data is stored.

Because the pointer points to the memory locations, the function can directly access the data in memory, and directly change the data in memory. Although this sounds like it could prove dangerous, with the proper use, this can be a lifesaver in your sketches.

Let's take a look at an example of passing pointers to a function and see just how that works.

▼ TRY IT YOURSELF

Passing Pointers to Functions

In this example, you define a function that accepts two pointers as parameters. The function will calculate the addition and subtraction results of two values and make the results of both calculations available to the calling sketch code. Here are the steps to test this:

1. Open the Arduino IDE, and enter this code into the editor window:

```
void setup() {
   Serial.begin(9600);
   int a = 5;
   int b = 3;
   Serial.print("The intitial values are ");
   Serial.print(a);
   Serial.print(" and ");
   Serial.println(b);
   addsub(&a, &b);
   Serial.print("The addition result is ");
   Serial.println(a);
   Serial.print("The subtraction result is ");
   Serial.println(b);
}

void addsub(int *x, int *y) {
   int add = *x + *y;
   int sub = *x - *y;
   *x = add;
   *y = sub;
}

void loop() {
}
```

2. Save the sketch code as **sketch1105**.
3. Click the Upload icon to verify, compile, and upload the sketch to your Arduino unit.
4. Start the serial monitor to run the sketch and view the output.

When you run the sketch, you should see the output shown in Figure 11.5.

The first thing you should notice is that we passed the values to the `addsub` function as memory locations using the reference operator:

```
addsub(&a, &b);
```

Because we're passing the values as memory locations, we need to define the `addsub` function parameters as pointers:

```
void addsub(int *x, int *y) {
```

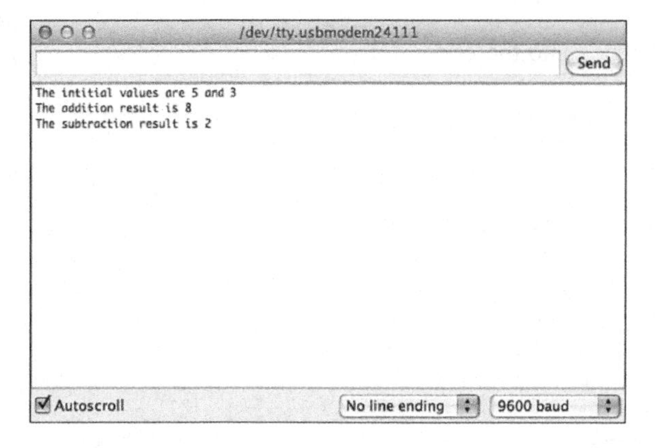

FIGURE 11.5
Using pointers to pass and retrieve values from functions.

Now that the function has the memory locations of both variables, it has to handle them as memory locations instead of data values. When we perform calculations, we need to use the dereference operator to retrieve the data values stored in the memory locations:

```
int add = *x + *y;
int sub = *x - *y;
```

And when we return the results, we need to place those values back in the memory locations:

```
*x = add;
*y = sub;
```

Notice that the code places the addition result into the memory location passed as the first parameter, and the subtraction result into the memory location passed as the second parameter. When using pointers, it's important to keep straight which memory locations will contain which return values.

WARNING

Returning Values in Pointers

It's also important to remember that when using pointers to return values from functions, the original values stored in the variables will be overwritten by the return values. If you need to retain the original values passed to the function, you must store them in another variable.

Summary

This hour covered the complicated topic of C data pointers. Pointers allow you to directly access the memory location where you store data on the Arduino, which can be a very powerful tool, but can also cause problems if you're not careful. The reference operator (&) allows you to retrieve the memory location where a variable is stored in memory, and the dereference operator (*) allows you to retrieve the value stored in a specific memory location. You must define the data type of a pointer so that the Arduino knows what type of data the pointer points to. That allows you to use pointer arithmetic to increase or decrease a pointer to point to additional data elements. Finally, you saw how to use pointers to retrieve more than one data value from a function, by passing memory locations to the function and allowing it to alter the data stored in those memory locations.

The next hour stays on the topic of storing data in memory on the Arduino and covers the different types of memory available and how to store your data in a specific type of memory.

Workshop

Quiz

1. What operator would you use to find the memory location where the variable named test is stored?

   A. The dereference operator `*test`

   B. The reference operator `&test`

   C. The incrementor operator `test++`

   D. The decrementor operator `test--`

2. The dereference operator returns the value stored at the memory location specified by the variable. True or false?

3. How can you tell whether a pointer assignment returns a valid memory address?

Answers

1. B. The reference operator returns the memory location where the Arduino stores the variable.

2. True. The dereference operator reads the value stored in the variable and then goes to that memory location and retrieves the value stored at that location.

3. You can assign a NULL value to the pointer before the assignment statement and then check whether the pointer is still NULL:

```
int *ptr = NULL;
ptr = &test2;
if (ptr) {
   Serial.println("Sorry, the pointer assignment failed");
}
```

Q&A

Q. **Is there a benefit to using pointers rather than arrays?**

A. With pointers, you can use pointer arithmetic to iterate through the stored numbers, whereas you can't do that with an array. That may save some coding from having to iterate through the array indexes.

Q. **Is there a benefit to using arrays instead of pointers?**

A. The one thing that pointers cannot do is return the number of data elements stored in the memory block. You can use the sizeof function on an array variable to return the number of data elements contained in the array, but that doesn't work with a pointer variable.

Q. **So which is better, pointers or arrays?**

A. Both have their use in different applications. You might find that you can use pointers to help cut down on your code if you have to do a lot of manipulation within the array of numbers. However, if you find yourself having to determine the number of values stored, you'll have to use arrays rather than pointers.

HOUR 12
Storing Data

What You'll Learn in This Hour:

- How the Arduino handles data in your sketches
- How to get the most use out of Arduino memory
- How to store data long term on an Arduino

One of the challenges of programming for the Arduino is that your sketches are limited by the resources available on the microcontroller. Nowhere is that more evident than when you try to handle large amounts of data. Because of the limited memory resources on the Arduino, you sometimes have to get a little creative in how you handle the data in your sketch. This hour shows some ways to help conserve memory in your sketches and shows how to utilize the extra EEPROM memory to store data for long-term use.

Arduino Memory Refresher

Hour 1, "Introduction to the Arduino," showed the basics of how the Arduino memory structure works. The ATmega AVR microcontroller chips used in the Arduino family have three different types of built-in memory:

- Flash
- Static random-access memory (SRAM)
- Electronically erasable programmable read-only memory (EEPROM)

The flash memory is where the Arduino stores the sketch executable code. Once you load your sketch into flash memory, it will remain there, even after you power off the Arduino. By default, the Arduino also installs a small bootloader program in flash memory, which assists in loading and starting your sketch each time you power on the Arduino.

The SRAM memory is where the Arduino stores the data used in your sketches. This is where any variable values that you define in your sketch are stored. The Arduino can access data in the SRAM memory area very quickly, but the downside is that the SRAM memory loses data when you turn off the power to the Arduino.

The EEPROM memory provides a long-term data storage solution for your Arduino sketches. Like the flash memory area, it retains any data that you store in the memory area after you remove power from the Arduino. This provides an area where you can store data that's used between sessions.

However, there are a couple of limitations to utilizing the EEPROM memory area on the Arduino. First, storing data in EEPROM is a relatively slow process, compared to SRAM data access speeds. That can significantly slow down your sketch.

The second limitation is that there are a limited number of times you can write to an EEPROM before the stored data become unreliable. The general rule of thumb is that after about 100,000 writes to the EEPROM, there's no guarantee that the data you write will get properly stored.

Table 12.1 shows the different sizes of flash, SRAM, and EEPROM available in the different Arduino units, based on the Atmel chip that they use.

TABLE 12.1 Arduino Memory Sizes

Chip	Flash	SRAM	EEPROM	Arduinos
ATmega168	16KB	1KB	512B	Pro, Nano, LilyPad
ATmega328	32KB	2KB	1KB	Uno, Mini, Fio, Ethernet
ATmega32u4	32KB	2.5KB	1KB	Leonardo, Yun, Micro, Esplora, Robot
ATMega2560	256KB	8KB	4KB	Mega
AT91SAM3X8E	512KB	96KB	0	Due

For the standard Arduino Uno, you only have 2KB of SRAM memory to work with in your sketches. The next section takes a closer look at how the Arduino manages your sketch data in the SRAM memory area.

DID YOU KNOW?

External Memory

The Arduino also allows you to add external memory to your project. The most common type of external memory to use is adding another EEPROM chip for storing data. This can come in handy if your project requires more data storage than what's available on the Arduino unit.

Taking a Closer Look at SRAM

The SRAM memory is the workhorse of the Arduino. It's where your sketch stores all the data that it uses as it runs. The CPU in the Arduino microcontroller handles memory management on the Arduino automatically for us, but you can use some tricks to help it get the most out of your SRAM memory area.

The ATmega AVR microcontroller CPU utilizes a common two-tier approach to managing the variables that you create in your sketch. This two-tier method divides the SRAM memory into two separate areas:

- The heap data area
- The stack data area

Each of these memory data areas handle different types of data that you define in the sketch. Figure 12.1 shows how the Arduino positions the heap and stack data areas within SRAM.

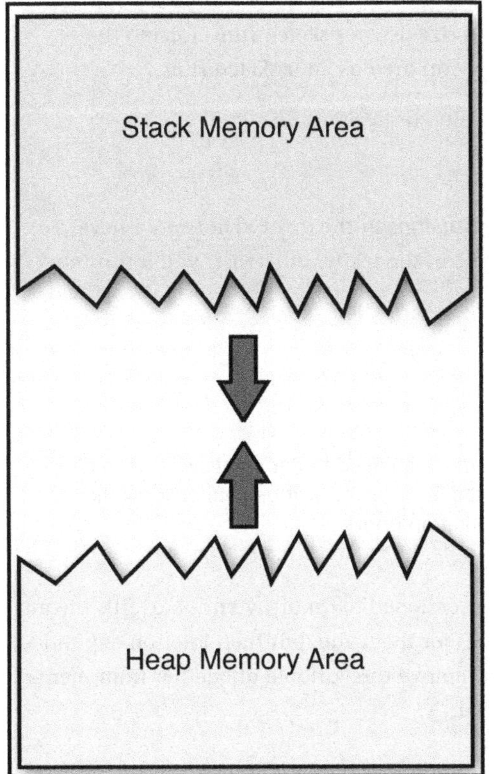

FIGURE 12.1
The Arduino SRAM heap and stack data areas.

The Arduino places the heap data area at the start of the SRAM memory area, and builds it upward, while it places the stack data area at the end of the SRAM memory area and builds it downward. The next sections take a closer look at each of these data areas in SRAM memory.

The Heap Data Area

The heap data area contains three different types of data that you define in your sketches:

- Global variables
- Variables defined with the `const` keyword
- Dynamic variables

Hour 5, "Learning the Basics of C," discussed the first two types of data. You define both global variables and constant variables at the beginning of your sketch, before you define your functions. When you define global and constant variables, the Arduino assigns them memory locations within the heap area in the lower portion of the SRAM memory block. These variables don't change in size, so they always take up the same amount of space in the heap as your sketch runs.

Dynamic variables are different. They can change in size as your sketch runs, forcing the Arduino to require more or less memory within the heap area as your sketch runs.

With static variables, you declare an integer array using a specific size:

```
int test[5];
```

And that's how much space in memory the Arduino assigns to the array. That space never changes in size. You can't store more than five values in the array; otherwise, you'll run into problems.

WARNING

Buffer Overflow

Trying to store more data than what's allocated in memory is called buffer overflow. This is dangerous in that the Arduino places variables next to each other in memory. If one variable overflows, it changes the value in another variable, without generating an error.

With dynamic variables, you can reallocate memory assigned to an array variable. This means you can start out defining space to reserve five values for the array, but then later on expand that to ten values, shrink it to three values, or even remove the variable altogether from memory. You'll see just how to do that later on in this hour.

Because the dynamic variables can change in size, the heap data area that stores them can change in size. The Arduino builds the heap "upward" in memory as the sketch requests more memory. If you remove or decrease the amount of memory assigned to a dynamic variable, the heap data area can become fragmented, as empty space appears within the heap memory area. The Arduino CPU attempts to store any new data requests in the fragmented areas first before growing the heap area.

The Stack Data Area

The stack data area is where the Arduino stores any local variables that you declare in functions, as well as internal pointers used to keep track of functions within the sketch. The odd thing about the stack is that the Arduino starts placing data in the stack starting at the end of the SRAM memory area, and builds it "downward" in memory.

When you call a function in your sketch, the Arduino stores the local variables that the function creates at the bottom of the stack area, and then removes them when the function finishes processing. This causes the stack area to grow and shrink frequently as your sketch runs.

So to sum things up, the heap data area can dynamically grow upward in memory, and the stack data area can dynamically grow downward in memory. As you can guess, problems arise if the stack area and heap grow large enough that they meet.

This is called a *memory collision*, and will result in corrupt data values in memory. If a collision occurs between the heap and stack data areas, you won't get an error message from your sketch, but you will start getting strange values in your sketch variables. As you can imagine, this can be a dangerous situation and cause all sorts of odd problems in your sketch.

The idea behind proper sketch data management is to try to avoid memory collisions in SRAM memory. To do that, sometimes you have to use the other memory areas in the Arduino to store data values to help free up space in the SRAM memory.

Creating Dynamic Variables

The heap data area in memory allows us to create our own dynamic variables to use in our sketches. The downside to that benefit is that you are responsible for manually controlling the dynamic variables, the Arduino won't do that for us.

That means you must allocate the memory to use, reallocate more memory if required, and release the dynamic variable from memory when you're done. If you don't properly follow the steps to release the memory when your sketch is done, this will result in what's called a *memory leak*.

This section walks through creating a dynamic variable in the heap memory area, and then how to manipulate it as your sketch runs, and remove it when you're finished.

Defining a Dynamic Variable

The C programming language provides two functions for dynamically allocating data for variables:

- `malloc`
- `calloc`

The `malloc` function defines a block of memory to assign to a variable, and then returns a void pointer to the start of the memory block. (If you remember from Hour 11, "Pointing to Data," a void pointer doesn't have a data type associated with it.)

BY THE WAY

Out of Memory

If the Arduino doesn't have enough unused SRAM memory available to satisfy the `malloc` request, it returns a `NULL` pointer value. You can easily check that using an `if-then` condition check to determine whether your sketch has used up all the memory.

As part of the `malloc` function call, you must provide a single parameter that specifies the size of the block of memory you want to try to reserve:

```
int *buffer;
buffer = (int *)malloc(10 * sizeof(int));
```

The first line simply defines a pointer for an integer value. The second line is somewhat complicated, but is easier if you break it down into parts. First, it uses a feature called *type-casting* to change the void pointer that the `malloc` function returns into an integer pointer so that you can use the standard pointer arithmetic with it.

The parameter passed to the `malloc` function uses the `sizeof` function to retrieve the size of memory the system uses to store a single integer value. This guarantees your sketch code will work correctly across multiple CPU types if necessary. By multiplying the result of the `sizeof` function by 10, the `malloc` function reserves enough memory space to store 10 integer values.

The `calloc` function is similar to the `malloc` function, but it includes a second parameter that specifies the number of blocks of memory to reserve:

```
int *buffer;
buffer = (int *)calloc(10, sizeof(int));
```

After running these statements, the Arduino will reserve a block of memory for storing 10 integer values and return the starting memory location in the `buffer` pointer variable. Now you can use the `buffer` pointer variable just like a normal array variable, with 10 data elements in the array. The difference is that you can change the number of data elements assigned to the array, as the next section shows.

Altering a Dynamic Variable

After you reserve a block of memory in the SRAM heap area using the `malloc` or `calloc` functions, you can dynamically change that using the `realloc` function. The `realloc` function takes two parameters, the original buffer that points to the dynamic memory area, and the size you want for the new area:

```
buffer = (int *)realloc(buffer, 20 * sizeof(int));
```

This statement dynamically changes the `buffer` variable memory block to a size of 20 and then reassigns the pointer to the `buffer` variable. Just as with the `malloc` and `calloc` functions, if the memory request fails the `realloc` function returns a `NULL` pointer value.

Removing a Dynamic Variable

The downside to using dynamic variables is that it's your responsibility to remove them from memory when you're done using them. If you don't, the Arduino CPU won't know that they're finished and will keep them in memory for as long as the system is running. This is what causes memory leaks in sketches.

To remove a dynamic variable from the heap data area, you use the `free` function:

```
free(buffer);
```

After you release the memory area, you can no longer use the buffer variable in your sketch to reference the memory location. You won't get an error message if you do, but it will no longer point to a reserved area in memory, and the Arduino may well have already placed other data in the same location.

Putting It All Together

Let's go through an example that demonstrates how to use a dynamic variable in a sketch to see just how all of these dynamic memory functions work.

TRY IT YOURSELF ▼

Creating and Using Dynamic Variables

In this example, you create a dynamic array variable in the heap data area, assign some data values to it, and then change the size of the array so that you can assign more data values to it. To run the example, just follow these steps:

1. Open the Arduino IDE, and enter this code into the editor window:

```
void setup() {
   Serial.begin(9600);
   int test[5] = {1, 2, 3, 4, 5};
   int i;
   int *buffer;
   buffer = (int *)calloc(5, sizeof(int));
   char output[50];
   sprintf(output, "The static array variable is stored at %x", &test[0]);
   Serial.println(output);
   for(i = 0; i < 5; i++) {
      buffer[i] = i;
      sprintf(output, "The dynamic value stored at %x", &buffer[i]);
      Serial.print(output);
      Serial.print(" is ");
      Serial.println(buffer[i]);
   }
   Serial.println("Now allocating more memory..");
   buffer = (int *)realloc(buffer, 10 * sizeof(int));
   if (buffer == NULL) {
      Serial.println("Unable to allocate more memory");
      exit(1);
   } else {
      Serial.println("Successfully allocated more memory");
      for(i = 5; i < 10; i++) {
         buffer[i] = i;
         sprintf(output, "The dynamic value stored at %x", &buffer[i]);
         Serial.print(output);
         Serial.print(" is ");
         Serial.println(buffer[i]);
      }
   Serial.println("Freeing the memory...");
   free(buffer);
}
}
void loop() {
}
```

2. Save the sketch code as **sketch1201**.
3. Click the Upload icon to verify, compile, and upload the sketch to your Arduino unit.
4. Open the serial monitor utility to run the sketch and view the output in the serial monitor window.

The sketch1201 code creates a dynamic array pointer variable called `buffer` that reserves memory to store 5 integer values, stores some data in them using the `buffer` pointer, and then uses the `realloc` function to request memory to store 10 integer values and store data in them. Figure 12.2 shows the output you should see in your serial monitor. (The memory locations may differ for your specific Arduino unit.)

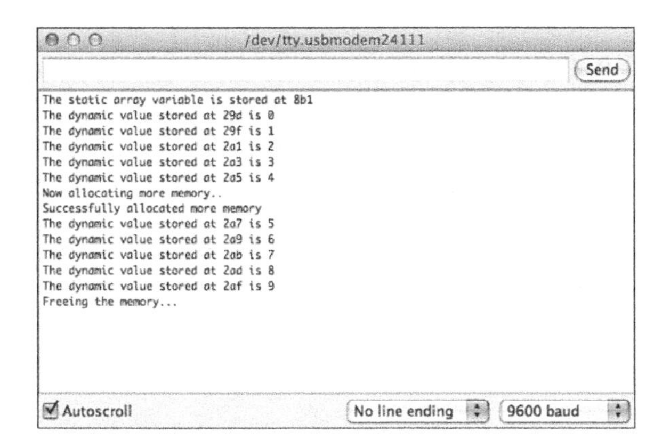

FIGURE 12.2
Using `malloc` and `realloc` to dynamically assign variables in memory.

The sketch also creates a standard static integer array variable called test and then displays the memory location where the Arduino stores it. This enables you to compare the locations where the Arduino stored the static and dynamic variables. Notice that the static variable's memory location is higher in memory than the dynamic variable's location. That's because the static variable is stored in the stack data area, while the dynamic variable is stored in the heap data area.

The dynamic variable feature available in the heap can come in handy as you create your sketches. For example, if you're not sure just how many data points you'll retrieve from a sensor, you can dynamically alter your array to store them all on-the-fly.

Using Flash to Store Data

As you saw in Table 12.1, the SRAM memory area on the Arduino is somewhat limited in size. If you have to write large sketches that use lots of data, you may bump up against the memory limit and start getting memory collisions. To help solve this problem, you can free up some space in SRAM by utilizing the flash memory area to store some of the sketch data values.

While the flash memory area is primarily reserved for storing your sketch code, you can also use it to store static data values in your sketch. Because the flash memory area is only written to when you upload your sketch code, you can't store any data that dynamically changes as your sketch runs, because the sketch code can't overwrite the flash memory area itself.

This feature can come in handy for storing constant values required in your sketch, such as any text messages that you use for output. This section walks through the steps required to store and use data in the flash memory area.

Declaring the Flash Data Types

Storing your sketch data in the flash memory area requires using some special data types and functions built specially for the Arduino. Because these aren't standard C programming features, you'll have to load a special code library file into your sketch.

Hour 13, "Using Libraries," discusses how to use external library files in your sketch, but basically, all you'll need to do to load the Arduino flash library data types and functions is add this line to the start of your sketch:

```
#include <avr/pgmspace.h>
```

After you load the library, you can access the special data types and functions required to interface with flash memory. Table 12.2 shows the special data types that you must use for storing data in the flash memory area.

TABLE 12.2 Flash Memory Data Types

Data Type	Bytes	Description
prog_char	1	A signed character
prog_int16_t	2	A signed integer
prog_int32_t	4	A signed long
prog_uchar	1	An unsigned character
prog_uint16_t	2	An unsigned integer
prog_uint32_t	4	An unsigned integer

At the time of this writing, there isn't a floating-point data type that you can use to store floating-point values in flash memory. You can only store integer and character values.

Also, to declare a variable that should be stored in flash memory, you must add the PROGMEM keyword to the variable declaration statement, like this:

```
prog_uint16_t maxTries PROGMEM = 10;
```

When you add the PROGMEM keyword to the variable declaration, it triggers the compiler to store the variable value in flash memory. Remember, after you do that, you cannot change the value assigned to the variable, it must remain constant.

Similarly, you can use this method to create constant a character array string value in your sketches:

```
prog_uchar message1[] PROGMEM = "This is a test message";
```

This can really come in handy if your sketch must store a lot of text messages for output. By placing the string values in flash memory, you free up extra space in the SRAM memory heap data area.

The second part of the process is accessing the data that you store in the flash memory area. Unfortunately, that's not as straightforward as just referencing the variables. The next section shows how to do that.

Retrieving Data from Flash

The downside to storing data in flash memory is that you can't access the data directly using the standard C functions. Because the flash memory is a separate memory space, they require special functions to access them. The pgmspace.h library provided by AVR includes several functions that allow us to access data stored in the Flash memory area, shown in Table 12.3.

TABLE 12.3 Functions to Access Flash Memory

Function	Description
`pgm_read_byte(`*location*`)`	Read 1 byte from flash memory.
`pgm_read_word(`*location*`)`	Read a 2-byte word from memory.
`strcmp_P(`*local, flash*`)`	Compare a string in flash to one in SRAM memory.
`trcpy_P(`*local, flash*`)`	Copy a string in flash to one in SRAM memory.
`strlen_P(`*flash*`)`	Return the size of a string stored in flash.

The most common way to handle strings stored in flash is to copy them into a variable stored in SRAM memory when you need to use them. That way you can store all of the strings in flash, and only one variable in SRAM memory to use as a buffer, like this:

```
prog_char message1[] PROGMEM = "This is a test string";
char buffer[50];
strcpy_P(buffer, message1);
```

You can also get fancy by creating a table of the string pointers in flash memory:

```
prog_char *table[] PROGMEM = {message1, message2, message3};
```

You can then use the `pgm_read_word` function to read the appropriate pointer to access the strings stored in flash memory:

```
strcpy_P(buffer, (char *)pgm_read_word(&(table[0])));
```

Let's go through an example of how to do just that.

Testing It Out

Now that you've seen the individual pieces required to store and retrieve data in the flash memory area, let's walk through an example that demonstrates how to do that.

▼ TRY IT YOURSELF

Storing Strings in Flash Memory

In this example, you store a series of string values in flash memory so that they don't take up space in the SRAM memory area and then create a table of pointers that your sketch code uses to access the strings as needed. Here are the steps to create and run the demo sketch:

1. Open the Arduino IDE, and enter this code into the editor window:

```
#include <avr/pgmspace.h>

prog_char message1[] PROGMEM = "This is a short message";
prog_char message2[] PROGMEM = "This is a little longer message";
prog_char message3[] PROGMEM = "Test message 3";
prog_char message4[] PROGMEM = "This is the last message in the test";

prog_char *table[] PROGMEM = {message1, message2, message3, message4};

void setup() {
  Serial.begin(9600);
  char buffer[50];
  int size;
  for(int i = 0; i < 4; i++) {
     size = strlen_P((char *)pgm_read_word(&(table[i])));
     strcpy_P(buffer, (char *)pgm_read_word(&(table[i])));
     Serial.print("The size of the string is: ");
     Serial.print(size);
```

```
        Serial.print(": ");
        Serial.println(buffer);
    }
}
void loop() {
}
```

2. Save the sketch code as **sketch1202**.

3. Click the Upload icon to verify, compile, and upload the sketch to your Arduino unit.

4. Open the serial monitor to run the sketch and view the output.

The strings are defined as character arrays, except that they use the `prog_char` data type. The table also uses the `prog_char` pointer type and contains the pointers to each of the strings stored in the flash memory. The code uses the special `strlen_P` and `strcpy_P` functions to access the strings in flash memory, using the `pgm_read_word` function to read the pointer address to use for the reference. Figure 12.3 shows the output that you should see in the serial monitor window.

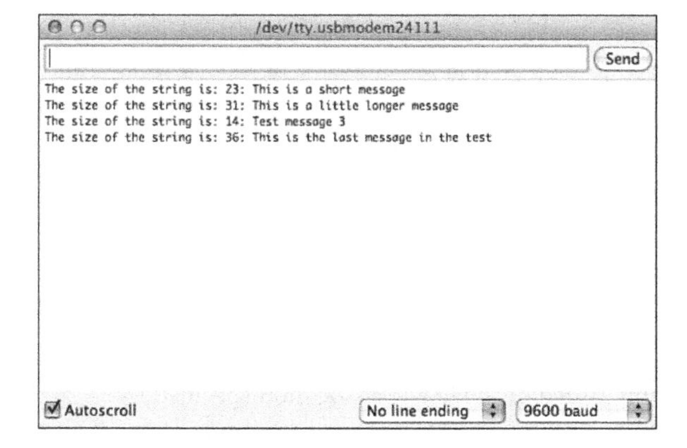

FIGURE 12.3
Storing strings in flash memory and retrieving them.

By using this method, you can store as many strings as you need in flash memory (or at least until you run out of flash memory), saving space in SRAM memory for your program variables.

Using the EEPROM Memory

The last trick to help with memory management in the Arduino is the EEPROM memory. With EEPROM memory, you can not only store and retrieve data outside of the SRAM memory area, but you can also do that in different runs of the sketch code. The EEPROM memory retains the data after a reboot of the Arduino (even after you power down the Arduino and restart it), making it ideal for storing values that your sketches need to "remember" from one run to another.

WATCH OUT

Writing to EEPROM

A limitation applies to the EEPROM memory. Just like the flash memory, it becomes unreliable after about 100,000 writes. You won't receive any error messages when the write data, it just may not complete properly. Keep this in mind when writing to your EEPROM memory area.

The following sections show how to use the Arduino EEPROM library to read and write data to the EEPROM memory area.

Using the EEPROM Library

To use the EEPROM memory area, you must use special functions. Just as with the code to access the flash memory area, you'll first need to include a library file in your sketch to use the EEPROM functions. The EEPROM library is part of the Arduino IDE standard package, so you can include the library file in your sketch by clicking Sketch > Import Library > EEPROM from the menu bar. This automatically adds the `#include` line to your sketch:

```
#include <EEPROM.h>
```

After you include the standard EEPROM library file, you can use the two standard EEPROM functions in your code:

- `read(`*`address`*`)`—to read the data value stored at the EEPROM location specified by *`address`*.
- `write(`*`address`*`, `*`value`*`)`—to read values to the EEPROM location specified by *`address`*.

Each address value starts at 0 for the start of the EEPROM memory area, and increases until the end of the EEPROM memory area. Depending on the Arduino unit you have, the end value will be different. The Arduino Uno unit contains 1KB of EEPROM memory. Because 1KB is actually 1024 bytes, the largest memory area you can write to on the Uno is 1023. (EEPROM memory starts at address 0.)

WATCH OUT

Memory Address Wrap

Unfortunately, the Arduino will allow you to write to a memory address larger than the amount of EEPROM memory on your Arduino unit. If the address value is larger, the Arduino wraps around in memory. So, if you have an Arduino Uno unit and write to memory address 1024, the data value is stored in address 0. Be careful when specifying the address values in your sketches!

Unfortunately, there are no automated ways to handle data stored in the EEPROM; you must keep track of all data you store in the EEPROM manually. It can help to draw up a memory map that diagrams what values you store where in the EEPROM memory area.

Experimenting with the EEPROM Library

Let's go through an example that stores values in EEPROM and then reads them back. First, we'll create a sketch that stores the values in the EEPROM memory area. Follow these steps:

1. Open the Arduino IDE, and select Sketch > Import Library > EEPROM.
2. Enter this code into the editor window under the `#include <EEPROM.h>` directive that was added by the library import:

```
#include <EEPROM.h>

void setup() {
   Serial.begin(9600);
   Serial.println("Storing data in EEPROM...");
   int test[] = {10, 20, 30, 40, 50};
   int i;
   for(i = 0; i < 5; i++) {
      EEPROM.write(i, test[i]);
   }
   Serial.println("Data written to EEPROM");
}

void loop() {
}
```

3. Save the sketch code as **sketch1203**.
4. Click the Upload icon to verify, compile, and upload the sketch code to your Arduino unit.
5. Open the serial monitor to run the sketch code and view the output.

The sketch1203 code uses the write function from the EEPROM library to store the array values in the first five spaces in EEPROM memory. You should see the output shown in Figure 12.4 in the serial monitor.

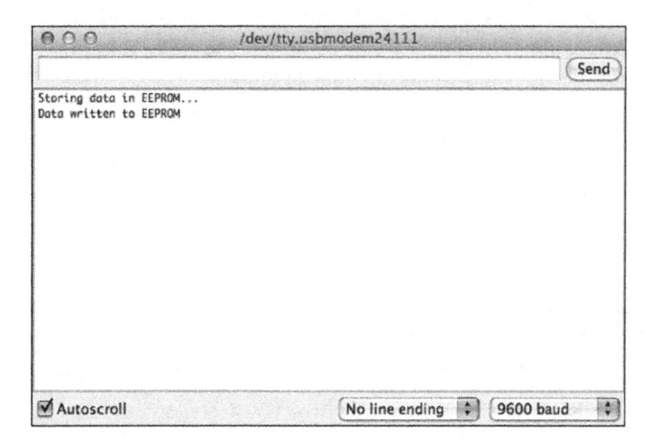

FIGURE 12.4
Storing integer values in EEPROM memory.

The next set of steps creates a second sketch to read the values that you stored in the EEPROM memory:

1. Open the Arduino IDE and select Sketch > Import Library > EEPROM.
2. Enter this code into the editor window under the `#include` directive that was added by the library import:

```
#include <EEPROM.h>

void setup() {
   Serial.begin(9600);
   int test2[5];
   int i;
   Serial.println("Reading data from EEPROM...");
   for(i = 0; i < 5; i++) {
      test2[i] = EEPROM.read(i);
      Serial.println(test2[i]);
   }
}

void loop() {
}
```

3. Save the sketch code as **sketch1204**.
4. Click the Upload icon to verify, compile, and upload the sketch code to your Arduino unit.
5. Open the serial monitor to run the sketch and view the output.
6. Disconnect your Arduino unit so that it loses power; then reconnect it back to your workstation.
7. Open the serial monitor again to run the sketch and view the output.

The output from the sketch1204 code should display the five data values that you stored from the sketch1203 code (as shown in Figure 12.5), even after removing power to the Arduino.

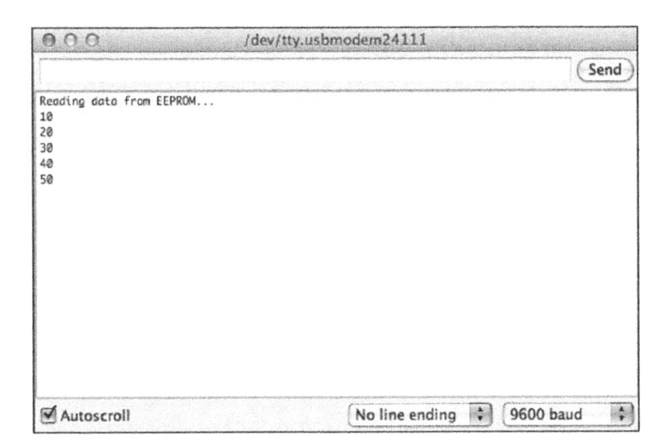

FIGURE 12.5
Retrieving data from the EEPROM memory area.

BY THE WAY

Storing More Complicated Data Structures

Because the EEPROM `read` and `write` functions can retrieve only 1 byte of data, they're somewhat limited in what they can store (only values from 0 to 255). However, some enterprising Arduino users have created a library of functions that allow you to easily store other data types, including character arrays and data structures, using special functions. This library is called the EEPROM Extended library, or EEPROMex for short. Unfortunately, it's not included as part of the standard Arduino IDE package, so you have to download and install it separately. You'll learn about how to do that in the next hour.

Summary

This hour focused on alternative ways to store data in your Arduino sketches. First, it showed just how the Arduino uses the SRAM memory area to store the variables that you declare in your sketches. After that, it showed how you can use the heap data area in SRAM to create and use dynamic variables that can change in size as your sketch runs. Next, it walked through how to store static values in flash memory to help free up space in SRAM. You must use special data types and functions to store data in the flash memory area with the program code. Finally, the hour discussed how to use the EEPROM memory that is built in to the Arduino CPU itself. You can use the standard EEPROM library to read and write individual bytes of data in EEPROM. This data remains intact even after the power is removed from the Arduino unit.

The next hour covers how to work with the different libraries available for the Arduino and how to create your own library of functions that you can share with others.

Workshop

Quiz

1. Where can you create dynamic variables that can change in size as you run your Arduino sketches?

    A. The SRAM stack data area

    B. The flash memory area

    C. The SRAM heap data area

    D. The EEPROM memory area

2. Data values stored in flash are lost when power is removed from the Arduino. True or false?

3. How do you retrieve a value stored in the EEPROM memory area?

Answers

1. C. You can create dynamic variables using the `malloc` or `calloc` functions in the SRAM heap data area.

2. False. The flash memory area retains the program and data information stored there after power is removed. When power is restored to the Arduino, the program runs, and the data values you stored in flash will also still be there.

3. You must use the `EEPROM.read` function available in the EEPROM library for the Arduino. You cannot use a standard C function to read or write data to the EEPROM memory area.

Q&A

Q. If the EEPROM memory area on my Arduino already contains data stored in it, can I read the data?

A. Yes, but you might not be able to make any sense out of it! You can read the data stored in the EEPROM 1 byte at a time, but you'd have to determine if the data stored was a character, number, or data structure. That could be close to impossible to figure out if you weren't the one who stored the data.

Q. Should I erase the EEPROM memory area by placing 0s (zeros) in all the memory locations when I'm done using it?

A. It depends on the sensitivity of your data. Remember, the more you write the EEPROM memory, the fewer times you have before it becomes unreliable. If there isn't a requirement to erase data from the EEPROM, I wouldn't bother doing it. You can overwrite old data with new data at any time.

HOUR 13
Using Libraries

What You'll Learn in This Hour:

- What an Arduino library is
- How to use standard Arduino libraries
- How to use contributed libraries
- Creating your own Arduino libraries

As you start writing larger Arduino sketches, you may find yourself reusing the same pieces of code for different sketches. In Hour 10, "Creating Functions," you saw how to create functions to help cut down on the amount of code you had to write for a single sketch, but if you wanted to share that code between multiple sketches, you'd still have to copy and paste it into each sketch. This hour shows you how to reuse functions without having to copy code by using libraries.

What Is a Library?

Libraries allow you to bundle related functions into a single file that the Arduino IDE can compile into your sketches. By compiling the library into your sketch, you can use any of the functions defined inside the library code anywhere in your sketch. Instead of having to rewrite the functions in your code, you just reference the library file from your code, and all the library functions become available.

This proves especially handy as you work with different Arduino shields. Each shield requires its own set of functions for accessing the hardware components on the shield, such as the network connection on the Ethernet shield, or the LCD display on an LCD shield. By bundling functions required for each shield into a separate library file, you can include only the libraries you need in your Arduino sketches.

There is a standard format that all Arduino libraries must follow for them to work correctly. The following sections discuss the standard Arduino library format.

Parts of a Library

Arduino libraries consist of two separate files:

- A header file
- A code file

The header file defines templates for any functions contained in the library. It doesn't contain the full code for the functions, just the template that defines the parameters required for the function, and the data type returned by the function.

For example, to define the template for a simple addition function that requires two integer parameters and returns an integer data type, the header file would contain the following line:

```
int addem(int, int);
```

This might look a little odd; the header template doesn't define any variables for the parameters, only the data types that are required. The compiler uses the header to determine the format of the functions as the sketch code uses them. That way it knows if you're not using a library function correctly, before it tries to compile the code.

The header file must use a .h file extension, and use the library name as its filename. Thus, if you want to call your library mymath, the header file would be mymath.h.

The code file in the library contains the actual code required to build the function. We'll take a more in-depth look at the code later on in this hour, but for now, it's only important to know that the code must be in C++ format. Because of that, the file extension you must use for the code body file is .cpp. So, for the mymath library, you must create the code file mymath.cpp.

Both files work together as the library. The header file helps the compiler know what format the sketch must use to call the library functions. That way it can flag an improperly used library function in your sketch code.

Library Location

The Arduino IDE must know how to access the library header and code files when it compiles your sketch. For that, you must place the library files into a specific folder within the Arduino IDE file structure.

The Arduino IDE stores the library files within its own Arduino folder structure. For Windows, the Arduino IDE installs under the Program Files folder. In OS X, it installs under the Applications/Arduino.app/Contents/Resources/Java/libraries folder. For Linux, there isn't a default location; the Arduino folder should be wherever you installed the software.

Under the Arduino application folder, you should see a folder named libraries. This is where the standard Arduino library files are stored. Each library is stored under a separate folder, and that folder name represents the name of the library.

BY THE WAY

Personal Libraries

Each user account on the workstation also has a personal library folder where you can place libraries that only you have access to use in the Arduino IDE. For both the Windows and OS X environments, look under the Documents folder for your user account for the Arduino folder. There is another library folder under there where you can store library files.

You can take a look at what libraries are included in your Arduino environment from the Arduino IDE menu bar. Just click the Sketch menu bar option, and then select the Import Libraries submenu option. By default, the Arduino IDE includes several standard libraries to support common shields that you can use with your Arduino unit. The next section discusses how to use the standard library files in your sketches.

Using the Standard Libraries

There are quite a few libraries that come standard with the base Arduino IDE installation. Table 13.1 lists the libraries that you'll find in the 1.0.5 version of the Arduino IDE.

TABLE 13.1 The Standard Arduino Libraries

Library	Description
EEPROM	Functions to read and write data to EEPROM memory.
Esplora	Functions for using the game features of the Esplora unit.
Ethernet	Functions for accessing networks using the Ethernet shield.
Firmata	Functions for communicating with a host computer.
GSM	Functions for connecting to mobile phone networks using the GSM shield.
LiquidCrystal	Functions for writing text to an LCD display.
Robot_Control	Functions for the Arduino Robot.
SD	Functions for reading and writing data on an SD card.
Servo	Functions for controlling a servo motor.
SoftwareSerial	Functions for communicating using a serial port.
SPI	Functions for communicating across the SPI port.

Library	Description
Stepper	Functions for using a stepper motor.
TFT	Functions for drawing using a TFT screen.
Wifi	Functions for accessing a wireless network interface.
Wire	Functions for communicating using the TWI or I2C interfaces.

As you can see from Table 13.1, there are lots of standard libraries associated with Arduino shield devices already available in the Arduino IDE. This makes working with shields a lot easier for the beginning developer. The next sections go through the steps you'll need to take to use library functions in your sketches.

Defining the Library in Your Sketch

To use a library in your Arduino sketch, just click the Sketch menu option from the menu bar, select the Import Library menu option, and then select the standard library you want to use from the list.

When you select a library, the Arduino IDE adds a line of code to your sketch, referencing the library header file. That line of code uses the `#include` directive (where *`libraryheader`* is the filename of the header file for the library):

```
#include <libraryheader>
```

Once you have that line in your sketch, you can start using the functions that are defined in the library.

Referencing the Library Functions

With the header file defined in your sketch, you can reference any of the functions contained in the library without getting a compiler error. However, you have to tell the compiler that the functions you're using are from a library.

To do that, you must specify the function using the library name as part of the function name:

```
Library.function()
```

For example, to use the `read` function from the `EEPROM` library, you use the following line:

```
EEPROM.read(0);
```

It's important to know the format for each function that you want to use in the library. Different functions require a different number and types of parameters. Fortunately, the Arduino developers have done an excellent job of documenting all the standard libraries and making that documentation easily available via the web at http://arduino.cc/en/Reference/Libraries.

Compiling the Library Functions

After you've created your sketch code using the library functions, you do not have to do anything special to compile it. The header and code body files in the library take care of all that work for us. Just click the Verify icon to verify and compile your code, or the Upload icon to verify, compile, and upload the sketch code to your Arduino unit.

TRY IT YOURSELF ▼

Using Libraries in Your Sketch

This example uses functions from the EEPROM standard library to store data in the EEPROM memory area and then retrieve it. Just follow these steps to run the experiment:

1. Open the Arduino IDE.
2. Click the Sketch option from the menu bar, select Import Library, and then select the EEPROM entry from the list. This places the `#include` directive at the top of your editor window, pointing to the EEPROM.h header file.
3. Enter the following code in the editor window, under the `#include` directive:

```
#include <EEPROM.h>

void setup() {
   Serial.begin(9600);
   int i, result;
   int test[5] = {1, 2, 3, 4, 5};
   Serial.println("Saving data in the EEPROM...");
   for(i = 0; i < 5; i++) {
      EEPROM.write(i, test[i]);
   }
   Serial.println("Retrieving data from the EEPROM...");
   for(i = 0; i < 5; i++) {
      result = EEPROM.read(i);
      Serial.print("Location ");
      Serial.print(i);
      Serial.print(" contains value: ");
      Serial.println(result);
   }
}

void loop() {
}
```

4. Save the sketch as **sketch1301**.
5. Click the Upload icon to verify, compile, and upload the sketch code to your Arduino unit.
6. Open the serial monitor to run the sketch and view the output.

The first for loop uses the EEPROM.write function to store values into the first five memory locations in the EEPROM memory area, and the second for loop uses the EEPROM.read function to read those values and display them. You should see the output shown in Figure 13.1 in your serial monitor to tell that the library code worked properly.

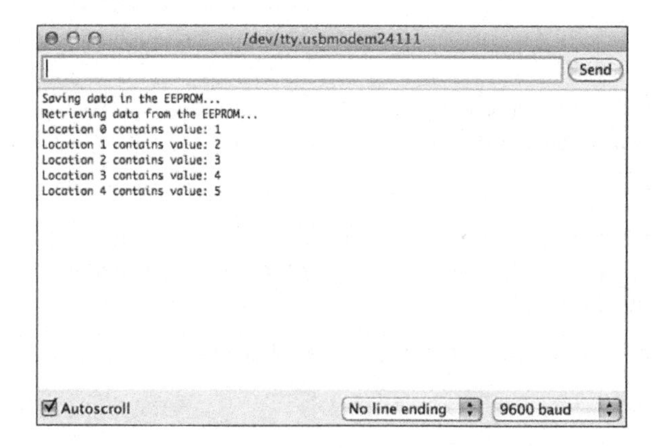

FIGURE 13.1
Using the EEPROM library code to access the EEPROM memory.

Using Contributed Libraries

Besides the standard libraries that the Arduino development group provides, Arduino users have created lots of other libraries. These are called *contributed libraries*.

One place to find a wealth of contributed libraries is the Arduino Playground website: http://playground.arduino.cc. From there, click the Libraries link on the left-side navigation bar to view a list of all the user-contributed libraries publicly available for the Arduino.

When you find a contributed library, you'll need to download it to your workstation and add it to your Arduino IDE libraries to be able to use it in your sketches. Fortunately, the Arduino IDE provides an easy interface for adding new libraries.

WATCH OUT

Multi-User Environments

If you have more than one user using the Arduino IDE on a single workstation, be careful, because the import library feature only imports the library for the current user. If other users on the workstation want to also use the library, they'll have to import it separately.

The following example shows the steps required to download and install the EEPROMex extended EEPROM library.

TRY IT YOURSELF 

Adding a Contributed Library

Once you find a contributed library that you want to use, you must install it into your Arduino IDE environment before you can use it. Here are the steps to do that:

1. Open a browser window and navigate to the http://playground.arduino.cc website.
2. From the main website page, click the Libraries link under the User Code Library section of the navigation bar on the left side of the web page. This should redirect you to this page: http://playground.arduino.cc//Main/LibraryList.
3. From the list of available libraries, scroll down to the Storage section and click the EEPROMex library link. This should redirect you to this page: http://playground.arduino.cc//Code/EEPROMex.
4. Click the link to download the library. Save the library zip file on your workstation in a location where you'll be able to easily find it.

WATCH OUT

Incorrect Link

Unfortunately, at the time of this writing, the URL link to the EEPROMex library is incorrect. Let's hope that gets fixed soon, but if not, the correct link is as follows:

http://thijs.elenbaas.net/downloads/?did=6

5. Close your browser window and start the Arduino IDE.
6. From the Arduino IDE, click Sketch > Import Libraries > Add Library. This produces a navigation dialog box.
7. In the resulting dialog box, navigate to the library zip file that you downloaded, select it, and then click the Open button.

8. The contributed library will be added to your installed libraries. Select Sketch; then Import Libraries from the menu bar. You should see the EEPROMex library listed under the Contributed section.

After you import the contributed library, you can use the functions defined in it without your sketches. Just remember to import the library into your sketch before you try to use them.

Creating Your Own Libraries

Another place where libraries can come in handy is to create your own libraries for functions that you commonly use in your Arduino projects. That helps cut down on the copying and retyping of code between sketches. This section walks through the process of creating your own library, installing it into the Arduino IDE, and then using it in your sketches.

Building Your Library

To create your own library of functions, you'll need to create both the code file and the header file and then bundle them together into a zip file to add to the Arduino IDE library. It's usually easier to create the code file first so that you know the formats for the header file.

Creating the Code File

When you create the code file for your library, you must use a slightly different format from what you're used to using in your Arduino sketches. So far, we've been using the C programming language format to create our sketches. To create library files, you must use C++ programming language format.

The C++ language uses an object-oriented programming format to define code classes. A code class bundles the variables and functions into a single class file that is portable between programs.

To start out, your library file must have a `#include` directive to reference the Arduino.h header file, along with an `#include` directive to reference the library header file:

```
#include "Arduino.h"
#include "MyLib.h"
```

Notice that these `#include` directives don't use the < and > symbols, but instead use double quotes around the header file names. That's because the header files aren't part of the standard library.

Next, you'll need to define the functions in your library. The C++ format defines class methods similar to how we defined functions (see Hour 10, "Creating Functions"). With C++ classes, you just define the methods inside a class name in the definition:

```
Classname::method
```

For example, to define the method addem in your MyLib library, you use this format:

```
int MyLibClass::addem(int a, int b) {
   int result = a + b;
   return result;
}
```

You can define additional functions in the library by just adding them to the same file. Let's go through creating the code file to use for our MyLib library file.

TRY IT YOURSELF 

Creating a Code File

In this exercise, you create the code file for our demo library. Just follow these steps:

1. Open a standard text editor on your system (such as Notepad in Windows or TextEdit on OS X). We can't use the Arduino IDE editor because we don't want to save the code as a sketch.
2. Enter this code into the text editor window:

```
#include "Arduino.h"
#include "MyLib.h"

int MyLibClass::addem(int a, int b) {
   int result = a + b;
   return result;
}

int MyLibClass::subem(int a, int b) {
   int result = a - b;
   return result;
}

int MyLibClass::multem(int a, int b) {
   int result = a * b;
   return result;
}

MyLibClass MyLib;
```

3. Save the file to your desktop (or some other common folder) as **MyLib.cpp**.

WATCH OUT

Text Files and Filenames

Be careful: Some text editors (such as Notepad) like to automatically append a .txt file extension to filenames when you save them. To get around that in Notepad, use double quotes around the filename when you save the file. For TextEdit on OS X, use the Preferences to select the default file type as Plain Text.

Besides defining the individual functions, the code declares an instance of the class that can be used by the sketch:

```
MyLibClass MyLib;
```

This is what allows you to reference functions using the format `MyLib.addem` in your sketch code, instead of having to create an instance of the entire class itself.

Now that you have the code file for the library created, you can build the header file. That's shown in the next section.

Creating the Header File

The header file creates a template that defines the functions contained in your library. The tricky thing is that it uses the C++ format to define a class for the functions.

First, it uses a directive to test whether the header file is present when compiling the library code. That uses the `#ifndef` directive:

```
#ifndef MyLib_h
#define MyLib_h
  // function templates go here
#endif
```

The `#ifndef` directive uses a somewhat odd format to check for the `MyLib_h` declaration. This is used in C++ programs to know when a header file has been included in the code. If the `MyLib_h` declaration is not found the code uses the `#define` directive to define it, then defines the function templates. Finally, the entire package is ended with a `#endif` directive to close out the `#ifndef` directive block.

The code to actually define the library functions uses the `class` identifier to define the library class, along with the `public` and `private` keywords to define public and private functions:

```
class MyLibClass
{
   public:
      // define public functions here
```

```
   private:
      // define private functions here
};
extern MyLibClass MyLib;
```

Public functions are functions that execute and return values outside of the class definition, such as in your sketches. These are the functions you want to share from your library in your sketches. Private functions allow you to create the "behind the scenes" functions that your public functions use, but aren't accessible from your sketches.

Let's go through the steps for creating the header file for our MyLib library package.

TRY IT YOURSELF 

Creating the Library Code File

Let's create a library of simple math functions that we can import into our sketches. Follow these steps:

1. Open a standard text editor on your system (such as Notepad in Windows or TextEdit on OS X).
2. Enter this code into the text editor window:

```
#ifndef MyLib_h
#define MyLib_h

#include "Arduino.h"
class MyLibClass {
   public:
      int addem(int, int);
      int subem(int, int);
      int multem(int, int);
};
extern MyLibClass MyLib;
#endif
```

3. Save the file to your desktop (or some other common folder) as **MyLib.h**.

Now you have both of the library files necessary to import into the Arduino IDE to use your functions. The next step is to package them for importing.

Building the Zip File

The Arduino IDE imports library files as a single zip-formatted file. You can create zip files using the Compressed Folder option in Windows and copying the MyLib.cpp and MyLib.h files into it.

Creating the zip file on OS X is a little more involved. Here are the steps to do that:

1. Create a folder called **MyLib**.
2. Copy the MyLib.cpp and MyLib.h files into the MyLib folder.
3. Right-click the MyLib folder, and select Compress "MyLib" from the menu options.

This creates a file called MyLib.zip that you can now import into your Arduino IDE.

Installing Your Library

After you've created the zip file with your library files, you're ready to install it into the Arduino IDE.

▼ TRY IT YOURSELF

Installing Your Library

When you have your zip library file created, you can import it into your Arduino IDE by following these steps:

1. Open the Arduino IDE.
2. Select Sketch from the menu bar.
3. Select Import libraries from the submenu.
4. Select Add library from the list of menu options.
5. This produces a dialog box. Navigate to the location of your library zip file, select it, and then click the Open button.

You should now see your MyLib library appear under the list of Contributed libraries in the Import Libraries section of the Sketch menu. You're all set to start coding with your new library now!

Using Your Library

Once the MyLib library file is imported into your Arduino IDE, you can use the functions in it in any sketch that you create. Just like any other library, you'll have to import it into your sketch first, and then you can reference the functions that you defined in the library using the library name, for example:

```
result = MyLib.addem(i, j);
```

The Arduino IDE uses the template defined in the library header file to look up the format of the addem function in your library to ensure that you're using it properly.

Let's test that by building a sketch that uses all the MyLib library functions.

TRY IT YOURSELF 

Using Your Library

To use your library, you need to first import it into your sketch, and then you're ready to use the functions. Here's a quick example that demonstrates how to use your new library in a sketch:

1. Open the Arduino IDE.
2. Click the Sketch option from the menu bar, then select Import Library, and then select the MyLib entry from the list. This should add the #include directive to the editor window showing your MyLib.h header file.

BY THE WAY

Missing Library

If you don't see the MyLib library listed in the Imported Libraries, double-check to make sure that there's a MyLib folder under the libraries section in your Arduino folder path. If not, try to import your library file again.

3. Enter the following code into the editor window, under the #include directive:

```
#include <MyLib.h>

void setup() {
  Serial.begin(9600);
  int i = 5;
  int j = 3;
  int result;
  result = MyLib.addem(i, j);
  Serial.print("The result of adding them is ");
  Serial.println(result);
  result = MyLib.subem(i, j);
  Serial.print("The result of subtracting them is ");
  Serial.println(result);
  result = MyLib.multem(i, j);
  Serial.print("The result of multiplying them is ");
  Serial.println(result);
}

void loop() {
}
```

4. Save the file as **sketch1302**.
5. Click the Upload icon to verify, compile, and upload the sketch to your Arduino unit.
6. Open the serial monitor to run the sketch and view the output.

The sketch code imports the MyLib library and then uses the functions defined in it. Figure 13.2 shows the output you should see in your serial monitor window if all went well.

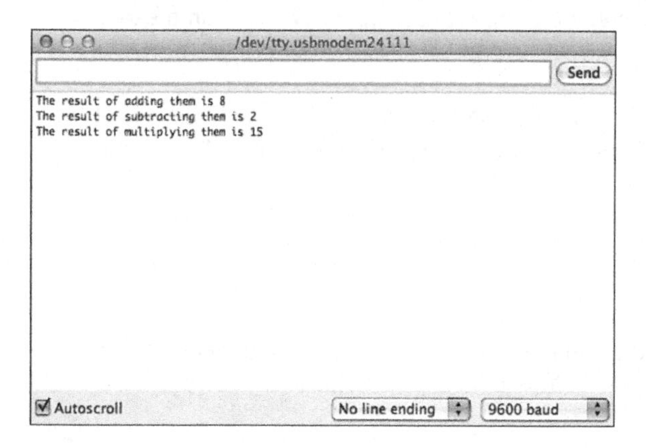

FIGURE 13.2
Output from using the MyLib library in your sketch.

You should now be comfortable with using libraries in your Arduino environment. As you go through the rest of the hours, there will be plenty of times when you'll need to import libraries to interact with the different shields.

Summary

This hour discussed how to use code libraries with the Arduino IDE. The Arduino IDE comes with some popular libraries already installed, making it easier to write code for the more popular Arduino shields. The hour also showed how to download libraries contributed by Arduino users from the Arduino playground website and import then into your Arduino IDE environment. You also saw how to create your own Arduino libraries with your own functions. That provides an easy way for you to share common code between all of your sketches without having to do a lot of extra typing.

The next hour starts looking at how to interface with the different hardware components of the Arduino, starting out with a look at how to work with the digital interfaces.

Workshop

Quiz

1. What file extension should you use for the code file in a library package?

   A. .ino

   B. .zip

   C. .h

   D. .cpp

2. When you import a library into the Arduino IDE, it's available for use for anyone that uses the workstation. True or false?

3. How do you bundle the library files so that they can be imported into the Arduino IDE?

Answers

1. D. The library code file must use the C++ code format, so it must use the .cpp file extension. You cannot use the standard Arduino .ino sketch file extension for code library files.

2. False. When you import a library into the Arduino IDE, it places the new library files in the user's local folder path. Other users on the workstation won't be able to access those library files. Each user who wants to use a contributed library must import the library on their own.

3. The library files must be enclosed in a compressed folder using the zip compression for the Arduino IDE to be able to import them.

Q&A

Q. Is there a limit to how many libraries I can include in a sketch?

A. Yes, each library takes up additional code space in flash memory when you compile and upload your sketch. The more libraries you include, the more room they take. Only include libraries that you know your code uses to help save space in flash memory.

Q. Do I have to bundle my library files in a zip file to import them?

A. Yes, the import utility expects the library files to be in a zipped file. However, you can manually create the library folder in your local Arduino library folder and copy the files manually if you prefer not to use the Arduino IDE import feature.

PART III

Arduino Applications

HOUR 14
Working with Digital Interfaces

What You'll Learn in This Hour:

- How the Arduino handles digital signals
- How to output a digital signal from a sketch
- How to read a digital signal in a sketch
- Different ways to detect digital inputs

Now that you've seen the basics of the Arduino programming language, it's time to start digging into the code to interface with the Arduino hardware. In this hour, you learn how to work with the digital interfaces on the Arduino, both reading digital data from external devices and sending digital signals out to them.

Digital Overview

The primary purpose of the Arduino is to control external devices. The Arduino provides two types of interfaces to do that:

- Analog interfaces for reading analog voltages
- Digital interfaces for reading and writing digital signals

This hour discusses the digital interfaces and focuses on the basics of how to read digital signals sent to the interface from an external device and how to write a digital signal to send to a device connected to an interface. In later hours, you'll see how to use those features to interact with specific sensors or other digital devices.

Each of the Arduino models provides a number of digital interfaces. Table 14.1 shows the number of digital interfaces available on each Arduino model.

TABLE 14.1 Arduino Digital Interfaces by Model

Model	Number of Digital Interfaces
Due	54
Leonardo	20
Mega	54
Micro	20
Mini	14
Uno	14
Yun	20

You can use each of the digital interfaces on the Arduino for either input or output. A few of the digital interfaces can also be used for other purposes (such as generating a pulse-width modulation signal or communicating with serial ports); later hours cover those topics. For now, let's just take a closer look at how to use the digital interfaces to interact with digital signals.

The Digital Interface Layout

The standard header layout in the Arduino footprint (see Hour 1, "Introduction to the Arduino") provides 14 digital interfaces along the top header block of the device, as shown in Figure 14.1.

FIGURE 14.1
Digital interfaces on the Arduino Uno.

The digital interfaces are labeled 0 through 13. This is the standard header layout available on all Arduino devices. For the Uno and Mini devices, those are the only digital interfaces available. For the Leonardo, Micro, and Yun models, the six analog interfaces, labeled A0 through A5, also double as digital interfaces 14 through 20.

For the Due and Mega models, a separate dual-row header block is included along the right side of the unit. The dual-row header block contains digital interfaces 21 through 53.

Setting Input or Output Modes

You can use each digital interface on the Arduino as either an input or an output, but not both at the same time. To tell the Arduino which mode your sketch uses for a specific digital interface, you must use the `pinMode` function:

```
pinMode(pin, MODE);
```

The `pinMode` function requires two parameters. The *`pin`* parameter determines the digital interface number to set. The *`MODE`* parameter determines whether the pin operates in input or output mode. There are actually three values you can use for the interface mode setting:

- **`INPUT`**—To set an interface for normal input mode.
- **`INPUT_PULLUP`**—To set an interface for input mode, but use an internal pullup resistor.
- **`OUTPUT`**—To set an interface for output mode.

The `INPUT_PULLUP` mode may be a bit confusing. Each Arduino model provides an option to activate an internal pullup resistor on each individual digital interface. The `INPUT_PULLUP` mode value determines whether the internal pullup resistor is activated on the digital interface or not. You'll learn more about that later on in this hour.

After you set the mode for the digital interface, you're ready to start using it. The next section discusses how to use the digital interfaces as outputs in your sketches.

Using Digital Outputs

For the output mode, your sketch can set the voltage level on each individual digital interface to either a logic `HIGH` or `LOW` value. You do that using the `digitalWrite` function:

```
digitalWrite(pin, value);
```

The *`pin`* parameter value specifies the digital interface number to send the output to. The *`value`* parameter specifies the `HIGH` or `LOW` setting to determine the output voltage. For example:

```
digitalWrite(10, HIGH);
```

The `HIGH` value sends a +5 volt signal to the digital output on interface 10. The Arduino generates digital signals using 40mA of current. For lower current devices, such as LEDs, you must place a resistor (typically around 1K ohms) in the circuit with the LED to help limit the current applied to the LED.

For high-current devices, such as motors and relays, you need to connect either a relay or a transistor to the digital output to use as a switch to control the high-current circuit. The transistor isolates the high-current circuit from the Arduino digital interface. When the digital interface is at a `HIGH` voltage level, the transistor conducts, allowing the higher-current circuit to flow. When the interface is at a `LOW` voltage level, the transistor blocks, breaking the higher-current circuit. Figure 14.2 demonstrates these configurations.

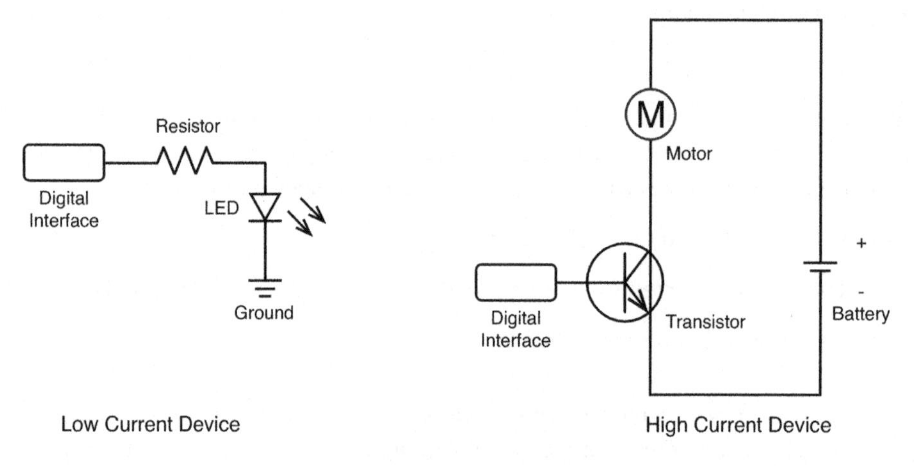

FIGURE 14.2
Connecting high- and low-current devices to the digital interface.

When connected directly to an electronic circuit, the digital interface on the Arduino can act as either a current source or as a current sink. When acting as a current source, the digital interface outputs a current when set to the `HIGH` voltage value. The electronic components connected to the Arduino interface must complete the circuit by providing a ground voltage (0 volts).

When acting as a current sink, the digital interface outputs a `LOW` voltage value of 0 volts, emulating a ground connection. In this case, the electronic circuit must provide current from a voltage source of +5 volts to complete the circuit. Figure 14.3 demonstrates both of these modes.

The next section demonstrates creating a sketch to show how to use the digital output features.

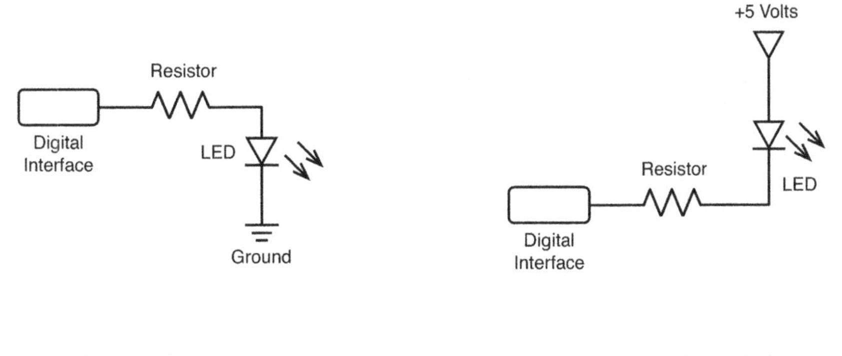

FIGURE 14.3
The Arduino digital interface as a current source or current sink.

Experimenting with Digital Output

Now that you've seen the basics of how to generate and control a digital signal from your Arduino, let's go through an example of using that feature with a real electronic circuit.

For this example, you need a few electronic components:

- Three LEDs (preferably red, yellow, and green, but they can be the same color if that's all you have available). Many Arduino starter kits come with a few 5mm, 30mA LEDs, and these will work just fine.
- Three 1K ohm resistors (color code brown, black, red).
- A breadboard.
- Four jumper wires.

The resistors are for limiting the current that goes through the LEDs. The Arduino can output 40mA of current, so if you use a 30mA LED, you'll need to place at least a 1K ohm resistor in series with each LED so it doesn't burn out. You can use a larger resistor if you like; the LED will just be a little dimmer.

After you gather these components, you can start the example.

▼ TRY IT YOURSELF

Digital Traffic Signal

For this example, you create a traffic signal that your Arduino will control using three separate digital interfaces. This project requires building a circuit, along with coding a sketch. First, follow these steps to build the electronic circuit:

1. Place the three LEDs on the breadboard so that the short leads are all on the same side and that the two leads straddle the space in the middle of the board so that they're not connected. Place them so that the red LED is at the top, the yellow LED in the middle, and the green LED is at the bottom of the row.
2. Connect a 1K ohm resistor between the short lead on each LED to a common rail area on the breadboard.
3. Connect a jumper wire from the common rail area on the breadboard to the GND interface on the Arduino.
4. Connect a jumper wire from the green LED long lead to digital interface 10 on the Arduino.
5. Connect a jumper wire from the yellow LED long lead to digital interface 11 on the Arduino.
6. Connect a jumper wire from the red LED long lead to digital interface 12 on the Arduino.

That completes the hardware circuit. Figure 14.4 shows the circuit diagram for what you just created.

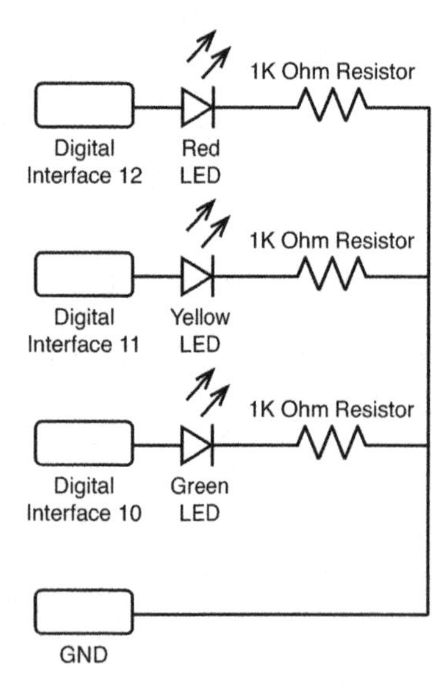

FIGURE 14.4
The circuit diagram for the traffic signal example.

Now you're ready to start coding the sketch that controls the traffic signal circuit. Just follow these steps:

1. Open the Arduino IDE on your workstation and enter this code into the editor window:

```
int stop = 6;
int yield = 2;
int go = 6;
void setup() {
   Serial.begin(9600);
   pinMode(10, OUTPUT);
   pinMode(11, OUTPUT);
   pinMode(12, OUTPUT);
   digitalWrite(10, LOW);
   digitalWrite(11, LOW);
   digitalWrite(12, LOW);
}

void loop() {
   stoplight(stop);
   golight(go);
   yieldlight(yield);
}

void stoplight(int time) {
   digitalWrite(10, LOW);
   digitalWrite(11, LOW);
   digitalWrite(12, HIGH);
   Serial.println("Light mode: Stop");
   delay(time * 1000);
}

void yieldlight(int time) {
   digitalWrite(10, LOW);
   digitalWrite(11, HIGH);
   digitalWrite(12, LOW);
   Serial.println("Light mode: Yield");
   delay(time * 1000);
}

void golight(int time) {
   digitalWrite(10, HIGH);
   digitalWrite(11, LOW);
   digitalWrite(12, LOW);
   Serial.print("Light mode: Go - ");
   Serial.println(time);
   delay(time * 1000);
}
```

2. Save the sketch as **sketch1401**.
3. Click the Upload icon to verify, compile, and upload the sketch code to your Arduino unit.
4. Open the serial monitor to start the sketch and to view the output from the sketch.

The `setup` function defines the pin modes for the three digital interfaces and sets their default values to `LOW`. This causes all the LEDs to remain off, because the LED circuits require a `HIGH` signal voltage for power.

Each state of the traffic signal (stop, go, and yield) is a separate function. Each function defines the power required to each LED for that state. (For example, in stop mode, interface 12 is set to `HIGH` to power the red LED, while interfaces 10 and 11 are set to `LOW` to turn off the yellow and green LEDs.) Also in each function is a `delay` function to pause the traffic light at that state for a predetermined amount of time, set by the value passed to the function. The global variables `stop`, `go`, and `yield` contain the number of seconds for each state.

When you run the sketch, the LEDs should light up simulating a U.S.-style traffic signal, and you should see an output log in the serial monitor window indicating which function is running at any given time.

BY THE WAY

Troubleshooting Interfaces

Using the serial monitor output is a great tool to help you troubleshoot digital interfaces. It helps you see inside the sketch to know when things should be happening. Once you get the traffic signal sketch working, you can remove the `Serial.println` function lines so that there isn't any logging output.

Working with Digital Inputs

For input mode, the digital interface detects a digital voltage provided by the external electronic device using the `digitalRead` function:

```
result = digitalRead(pin);
```

The *`pin`* parameter determines which digital interface to read. The `digitalRead` function returns a Boolean `HIGH` or `LOW` value based on the input signal, which you can compare to a 1 or 0 integer value in your code.

The digital interface can only detect binary digital signals, either a `true` value, or a `false` value. For the Arduino to detect a `true` value, the input voltage must be between +3 and +5 volts on the digital interface. For a `false` value, the Arduino must detect a voltage between 0 and +2 volts.

WATCH OUT!

Undetermined Values

Notice that this range leaves an area between +2 and +3 volts that is not assigned to a Boolean value. Any voltage detected by the digital interface within that range will produce an unreliable result.

The following sections cover a couple of issues you need to be aware of when using the digital interfaces as inputs.

Input Flapping

An issue often overlooked by novice Arduino users is input flapping. Input flapping occurs when a digital interface is not specifically connected to a source voltage (either +5 volts or ground). Figure 14.5 shows a circuit diagram that demonstrates this problem.

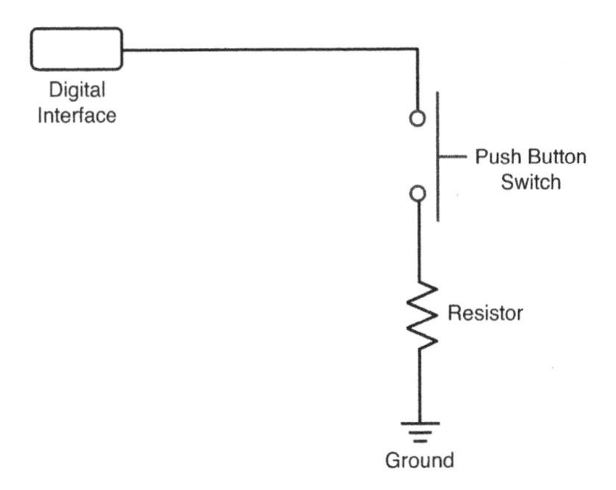

FIGURE 14.5
An Arduino interface that will experience input flapping.

When the switch in Figure 14.5 is depressed, the digital interface is connected to the ground voltage, which sets the input to a `LOW` value. However, when the switch is released, the digital interface is not connected to anything. When this happens, the `digitalRead` function returns an inconsistent value, often dependent on any ambient voltage that may be present on the interface. There's no guarantee that the input will return a `HIGH` value.

To prevent input flapping, you should always apply some signal to the digital interface at all times. You can do so in two ways:

- Connect the interface to +5 volts, called a pullup
- Connect the interface to ground, called a pulldown

In the pullup circuit, when the switch is open, the `digitalRead` function returns a `HIGH` value, and when the switch is depressed, it returns a `LOW` value. The pulldown circuit operates in the opposite way; when you depress the switch, the interface receives a `HIGH` value, but is at a `LOW` value otherwise. Figure 14.6 demonstrates this process.

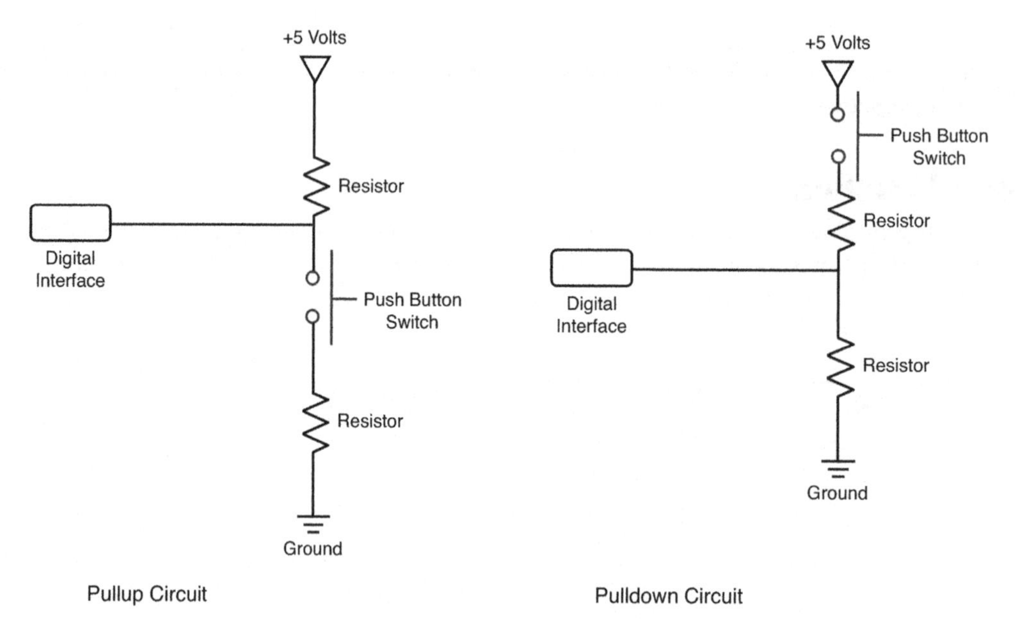

FIGURE 14.6
Using pullup and pulldown circuits on a digital interface.

Setting a pullup or pulldown circuit requires a little extra work on your part. Fortunately, another method enables you to accomplish that.

Using the Internal Pullup

Because of the importance of using a pullup or pulldown circuit, the Arduino developers provide an internal pullup circuit for each digital interface. You can activate the internal pullup circuit by using the `INPUT_PULLUP` label in the `pinMode` function:

```
pinMode(8, INPUT_PULLUP);
```

The `INPUT_PULLUP` label activates an internal pullup resistor on the specified digital interface, setting the default value to `HIGH` for the interface. The external circuit connected to the digital interface must connect the interface to ground to change the input voltage.

By using the internal pullup circuit, you don't need to create the external pullup circuit, thus reducing the amount of hardware you need for your projects.

Using Interface 13 as an Input

Digital interface 13 is somewhat special on all Arduino units. It has a resistor and LED connected to it at all times, allowing you to easily monitor output from that digital interface directly on the Arduino board. This comes in handy if you're trying to quickly troubleshoot a digital setting without having to connect external hardware.

The downside to this is that the LED and resistor are always connected to the interface. If you use interface 13 for input, any voltage applied to the interface will interact with the internal resistance applied to the input circuit. This means that the undetermined area between +2 and +3 volts becomes somewhat larger for that digital interface. To be safe, make sure to always apply a full +5 volts for a `HIGH` signal when using interface 13 for input.

Experimenting with Digital Input

Now that you've seen the basics of how to use digital interfaces for input, let's go through an example to demonstrate how that works. In this example, you add on to the traffic signal example you created earlier in this hour. You need three more components:

- A momentary contact switch
- A 10K ohm resistor (color code brown, black, orange)
- One jumper wire

The momentary contact switch is normally in an open state. When you press the button, the switch makes contact and is in a closed state, connecting the two pins of the switch to complete a circuit. When you release the button, the switch goes back to the open state.

The Arduino Starter Kit provides small momentary contact switches that you can plug directly into the breadboard. These work great for our breadboard examples. You can also purchase similar switches from most electronic parts suppliers, such as Adafruit or Newark Electronics.

TRY IT YOURSELF 

Using a Digital Input

In this example, you add a switch to your traffic signal circuit. When the switch is held closed, the traffic signal code increases the amount of time allotted for the green LED state. First, the steps you need to build the circuit:

1. Place the momentary contact switch on the breadboard under the existing LEDs so that the switch straddles the center divide of the breadboard.
2. Connect the 10K ohm resistor so that one lead connects to one pin of the momentary contact switch and the other lead connects to the breadboard rail connected to the Arduino GND interface.
3. Connect the jumper wire from digital interface 8 on the Arduino to the other pin of the momentary contact switch.

That's all you need to add for the hardware. Now, for the code changes, follow these steps:

1. Open the Arduino IDE.
2. Select File from the menu bar, and then select the sketch1401 code to open.
3. Add the `checkSwitch` function to the bottom of the existing sketch code:

```
int checkSwitch() {
  int set = digitalRead(8);
  Serial.print("checking switch...");
  if (set) {
     Serial.println("the switch is open");
     return 6;
  } else {
     Serial.println("the switch is closed");
     return 10;
  }
}
```

4. Modify the `setup` function to add a statement to set the interface mode for interface 8 to `INPUT_PULLUP`:

```
pinMode(8, INPUT_PULLUP);
```

5. Modify the `loop` function code to add a call to the `checkSwitch` function at the end of the loop, and assign the output to the go variable. Here's what the final `loop` function code should look like:

```
void loop() {
   stoplight(stop);
   golight(go);
   yieldlight(yield);
   go = checkSwitch();
}
```

6. Save the sketch code as **sketch1402**.
7. Click the Upload icon to verify, compile, and upload the sketch code to your Arduino unit.
8. Open the serial monitor to run the sketch and view the output.

The sketch1402 code uses the INPUT_PULLUP mode for interface 8 so that when the switch is not pressed, the input value will be set to HIGH. When you press the switch, the input will be set to LOW, as the switch connects the interface to the ground using the 10K ohm resistor. (The resistor helps prevent short-circuiting the digital interface if something goes wrong.) The checkSwitch function returns a different value based on whether the switch is pressed.

When you run the sketch, the default value for the go variable of 6 remains for as long as the switch is not pressed when the checkSwitch function runs. When you press and hold the switch while the checkSwitch function runs, the value for the go variable changes to 10.

The trick is to depress the switch while the checkSwitch function is running. To do that, just hold down the switch while the yellow LED is lit, and then release it when the red LED lights. When you do that, the next iteration of the green LED will be longer.

The output in the serial monitor will show the setting of the go variable in each loop iteration, as shown in Figure 14.7.

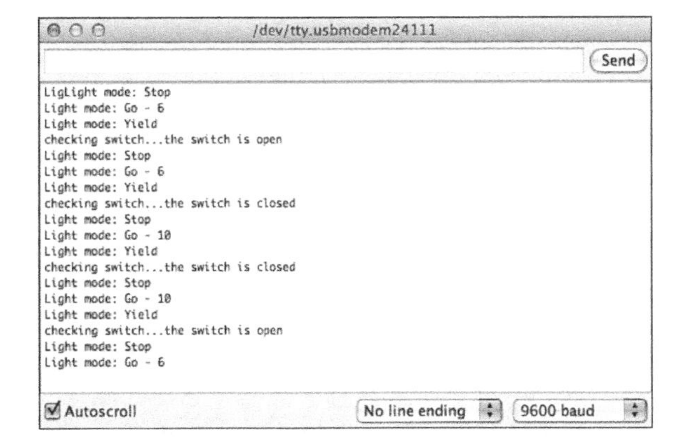

FIGURE 14.7
Output from controlling the traffic light using the switch.

When the switch is open, the green LED stays on for 6 seconds, but when you close the switch during the yellow LED time, the next iteration of the green LED stays on for 10 seconds.

Summary

This hour explored how to use the digital interfaces available on the Arduino. The Arduino provides multiple interfaces for detecting input digital signals, and also for generating output digital signals.

You must use the pinMode function to set the digital interface for input or output mode. For output signals, you use the digitalWrite function to generate a 40mA +5 volt output signal, which can power low-current devices. You can connect a transistor to the digital output to control higher-current devices.

For detecting digital input signals, you use the digitalRead function. You should always set a default value for the interface using either a pullup or pulldown voltage. You can do that either internally using the INPUT_PULLUP label in the pinMode function or externally by using the electronic circuit. This prevents input flapping, which will result in indeterminate values.

The next hour covers how to use the analog interfaces available on the Arduino. They enable you to both detect analog voltages for input and even to generate analog signals that you can use to control analog circuits, such as motors.

Workshop

Quiz

1. Which function do you use to activate the internal pullup resistor on a digital interface?

   A. digitalRead

   B. digitalWrite

   C. pinMode

   D. analogRead

2. If no voltage is applied to the digital interface, the digitalRead function will return a HIGH value. True or false?

3. What command should you use to set digital interface 3 to be used for input mode and to activate the internal pullup resistor?

Answers

1. C. The pinMode function allows you to use the INPUT_PULLUP mode to enable the internal pullup resistor on the specified digital interface.

2. False. If no input voltage is applied to the digital interface, the interface may flap between a HIGH and LOW value. You should always set an input interface to a specific voltage, either HIGH or LOW, at all times in your circuits.

3. `pinMode(3, INPUT_PULLUP);`

Q&A

Q. What happens if a voltage change occurs on the digital input interface while the sketch is doing something else and not running the `digitalRead` function?

A. The `digitalRead` function can only return the current value set on the digital interface at the time your sketch runs it. It can't detect whether a voltage change occurs between the times you run the command.

Q. How can I detect whether a change occurs between when the sketch runs the `digitalRead` function?

A. You can use interrupts to detect any changes on a digital interface. Those are discussed in Hour 16, "Adding Interrupts."

HOUR 15
Interfacing with Analog Devices

What You'll Learn in This Hour:

- How the Arduino handles analog signals
- How to read an analog signal
- How to output an analog signal
- Different ways to detect analog inputs

Besides working with digital devices, the Arduino can also interface with analog devices. However, working with analog signals can be a bit tricky because the digital Arduino must convert between the analog signal and digital values. Fortunately, the Arduino developers have made things easier for us by providing a group of simple functions to use. In this hour, you learn how to work with the Arduino analog functions to read analog signals from the analog input interfaces on the Arduino and how to generate analog output signals using the digital interfaces.

Analog Overview

Although microcontrollers are built around digital computers, often you'll find yourself having to interface with analog devices, such as controlling motors that require an analog input voltage or reading sensors that output an analog signal. Because of that, the Arduino developers decided to add some analog features to the Arduino microcontroller.

Working with analog signals requires some type of converter to change the analog signals into digital values that the microcontroller can work with, or changing digital values produced by the microcontroller into analog voltages that the analog devices can use. The next sections discuss how the Arduino handles analog input and output signals.

Detecting Analog Input Signals

For working with analog input signals, the Arduino includes an analog-to-digital converter (ADC). The ADC converts an analog input signal into a digital value. The digital value is scaled based on the value of the analog signal. Your Arduino sketch can read the digital value produced by the converter and use it to determine the value of the original analog signal.

Each of the Arduino units includes an ADC for sensing analog input voltages and converting them into digital values. Different Arduino models use different types and sizes of ADC converters. Table 15.1 shows the number of analog interfaces each Arduino model supports.

TABLE 15.1 Arduino Analog Interfaces by Model

Model	Number of Analog Interfaces
Due	12
Leonardo	12
Mega	16
Micro	12
Mini	8
Uno	6
Yun	12

The ADC converts the analog input voltage present on the interface to a digital value based on a preset algorithm. The range of digital values produced by the algorithm depends on the bits of resolution that the ADC uses.

For most Arduino models, the ADC uses a 10-bit resolution to produce the digital value. Thus, the digital values range from 0 for an input of 0 volts, to 1023, for an input of +5 volts. However, the Leonardo uses a 12-bit resolution ADC, which makes the upper value 4023 for +5 volts. This provides for a more granular result as the voltages change, allowing your sketch to detect smaller voltage changes on the input interface.

Generating Analog Output Signals

For generating analog signals for output, the converter is called a digital-to-analog converter (DAC). A DAC receives a digital value from the microcontroller and converts it to an analog voltage that is used to provide power to an analog device or circuit. The value of the analog voltage is determined by the digital value sent to the DAC; the larger the digital value, the larger the analog output voltage.

However, DAC devices are somewhat complicated and require extra circuitry to implement. Because of this, only one Arduino model, the Leonardo, includes an actual DAC to output true analog signals. If you own one of the other Arduino models, don't fret; there's another method for generating analog output signals.

Pulse-width modulation (PWM) simulates an analog output signal using a digital signal. It does that by controlling the amount of time a digital output toggles between the `HIGH` and `LOW` values. This is called the signal duty cycle.

The length of the signal duty cycle determines the simulated analog voltage generated by the digital interface. The longer the duty cycle, the higher the analog voltage that it simulates. A duty cycle of 100% means that the digital output is always at the `HIGH` value, which generates an output voltage of +5 volts. A signal duty cycle of 0% means that the digital output is always at the `LOW` value, which generates an output voltage of 0 volts.

For values in between, the digital output toggles between the `HIGH` and `LOW` values to produce a simulated analog signal. For example, when the duty cycle is 50%, the `HIGH` value is applied half of the time, which simulates an analog voltage of +2.5 volts. If the duty cycle is 75%, the `HIGH` value is applied three-fourths of the time, which simulates an analog voltage of +3.75 volts. Your sketch can use the duty cycle to simulate any analog voltage level from 0 volts to +5 volts on the Arduino.

Locating the Analog Interfaces

The standard header layout in the Arduino footprint (see Hour 1, "Introduction to the Arduino") provides six analog input interfaces along the bottom header block of the device, as shown in Figure 15.1.

The analog input interfaces on the Arduino are labeled A0 through A5. For the Uno and Mini devices, those are the only analog input interfaces available. For the Leonardo, Micro, and Yun devices, the additional analog interfaces are found on digital interfaces 4, 6, 8, 9, 10, and 12.

The Arduino only supports PWM output on a subset of the digital interfaces. For the Arduino Uno, digital interfaces 3, 5, 6, 9, 10, and 11 support PWM. You can tell which digital interfaces on your specific Arduino unit support PWM by the labeling on the unit. The PWM digital interfaces are marked on the Arduino using a tilde (~) before the digital interface number (see Figure 15.1).

FIGURE 15.1
Analog interfaces on the Arduino Uno.

Working with Analog Input

The main function that you use to work with the Arduino analog input interfaces is the `analogRead` function. It returns an integer value from the specified ADC interface, which represents the analog voltage applied to the interface:

```
int input;
input = analogRead(A0);
```

The analog interfaces use the `A` prefix to indicate the interface (A0 through A5 on the Uno). For most Arduino models, the `analogRead` function returns a digital value from 0 to 1023. For the Leonardo, the digital value is from 0 to 4023.

Let's go through an example to test the analog input on your Arduino unit.

▼ TRY IT YOURSELF

Detecting Analog input

In this example, you build a simple analog circuit that uses a potentiometer to change the input voltage on the Arduino analog input.

A potentiometer is a variable resistor that you control rotating a control arm. As you rotate the control arm, it changes the resistance between the pins of the potentiometer. The potentiometer has three pins. The two outer pins connect across the entire resistor in the potentiometer, providing a constant resistance value. The inner pin acts as a wiper; it scans across the resistor as you turn the control arm. The resistance value between one of the outer pins and the inner pin changes as you rotate the controller arm. The resistance generated ranges from 0 ohms to the maximum value of the potentiometer resistor. We'll use this feature to change the voltage present on the analog interface on the Arduino.

For this example, you need the following components:

- A potentiometer (any value)
- A breadboard
- Three jumper wires

Follow these steps to build the circuit for the example:

1. Plug the potentiometer into the breadboard so that the three pins are positioned across three separate rows.
2. Connect one outer pin of the potentiometer to the +5 pin on the Arduino header using a jumper wire.
3. Connect the other outer pin of the potentiometer to the GND pin on the Arduino header using a jumper wire.
4. Connect the middle pin of the potentiometer to the A0 pin on the Arduino header using a jumper wire.

This configuration allows you to change the resistance between the +5 volt source and the A0 analog interface on the Arduino as you turn the control arm of the potentiometer. Because the resistance changes, the amount of voltage applied to the interface will change. Figure 15.2 shows the circuit diagram for this.

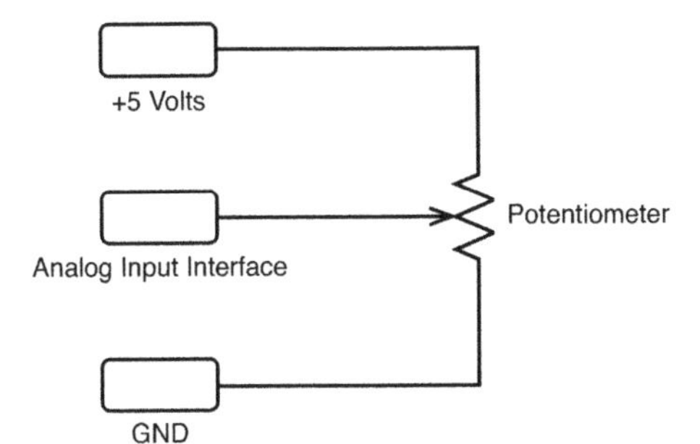

FIGURE 15.2
The analog input test circuit.

Next, you need to create a sketch to read the analog interface and display the retrieved value. To do that, follow these steps:

1. Open the Arduino IDE, and then enter this code into the editor window:

```
int input;
void setup() {
   Serial.begin(9600);
}

void loop() {
   input = analogRead(A0);
   Serial.print("The current value is ");
   Serial.println(input);
   delay(3000);
}
```

2. Save the sketch as **sketch1501**.
3. Click the Upload icon to verify, compile, and upload the sketch code to the Arduino unit.
4. Open the serial monitor to run the sketch and view the output.
5. As the sketch runs, rotate the potentiometer control arm and look at the values output in the serial monitor.

I placed a `delay` function at the end of the `loop` function so that the output doesn't continually stream to the serial monitor. As you rotate the potentiometer control arm, a delay will occur in the output that you see. You should see different values appear in the serial monitor output, as shown in Figure 15.3.

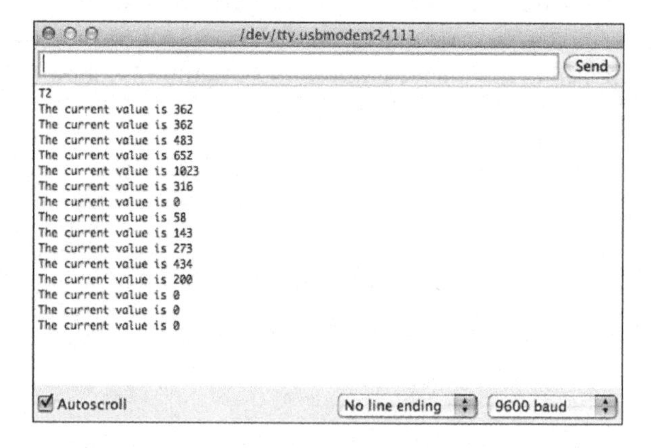

FIGURE 15.3
The Arduino output as you change the input voltage.

When you get to one end of the control arm rotation, the value returned by the `analogRead` function should be 0, indicating that no voltage is being applied to the analog input. When you get the opposite end of the control arm rotation, the value should be 1023, indicating that all 5 volts are being applied to the analog input.

Modifying the Input Result

By default, the `analogRead` function returns the detected voltage level as an integer value from 0 to 1023. It's up to your sketch to determine just what that value means. Fortunately, you can use a couple of functions in your sketches to help manage the values that you have to work with. This section covers those functions and shows how to use them.

Limiting the Input Values

You may run into a situation where you don't want your sketch to act on very low or very high values in the analog input signal range. For example, you may need a minimum voltage to control a motor, so your sketch needs to ignore any voltages detected on the analog input below a set threshold value.

To do that, you can use the `constrain` function:

```
constrain(value, min, max)
```

The *`min`* and *`max`* parameters define the minimum and maximum values returned by the `constrain` function. The *`value`* parameter is the value that you want tested by the `constrain` function. If the value is lower than the *`min`* value specified, the `constrain` function returns the *`min`* value. Likewise, if the value is larger than the *`max`* value specified, the `constrain` function returns the *`max`* value. For all other values between the *`min`* and *`max`* values, the `constrain` function returns the actual value.

You can test this out by adding the `constrain` function to the sketch1501 code that you used for the example:

```
void loop() {
   input = constrain(analogRead(A0), 250, 750);
   Serial.print("The current value is ");
   Serial.println(input);
   delay(3000);
}
```

Now as you run the program, you should see that the minimum value retrieved will be 250 instead of 0, and the maximum value retrieved will be 750 instead of 1023.

Mapping Input Values

You may have noticed that one downside to using the `constrain` function is that it makes for a large dead area below the minimum value and above the maximum value. For example, as you turn the potentiometer's control arm, the output value stays at 250 until the output actually gets above 250. This can prove impractical at times because it makes it more difficult to scale the input device to produce meaningful output values.

One solution to that problem is to use the `map` function:

```
map(value, fromMin, fromMax, toMin, toMax);
```

The `map` function has five parameters:

- ***value***—The value to scale.
- ***fromMin***—The minimum value in the original range.
- ***fromMax***—The maximum value in the original range.
- ***toMin***—The minimum value in the mapped range.
- ***toMax***—The maximum value in the mapped range.

In a nutshell, the `map` function can alter the range of a value from one scale to another. For example, if you want to change the range of the analog input to 0 through 255 rather than 0 through 1023, you use the following:

```
input = map(analodRead(A0), 0, 1023, 0, 255);
```

Now as you rotate the potentiometer control arm, the output will be between 0 and 255. You can test this out by adding the `map` function to the sketch1501 code that you used for the example:

```
void loop() {
   input = map(analogRead(A0), 0, 1023, 0, 255);
   Serial.print("The current value is ");
   Serial.println(input);
   delay(3000);
}
```

Now when you run the sketch, you'll notice that the values don't change as quickly as with the wider range. However, now you don't have a large dead spot before the minimum value or after the maximum value.

Using Input Mapping

One nice feature of the map function is that you can restrict the final values returned by the input to a predetermined result set. You can then use that result set in a switch statement to run code based on a range of the input values.

This example demonstrates how to do just that.

TRY IT YOURSELF ▼

Mapping Input Values

You can use the `map` function to map input values to any result set range that you need in your application. In this example, you map the input to one of three values (1 through 3), and then use that result to light a specific LED.

This example uses the same potentiometer circuit you created for the first example, plus it uses the traffic signal circuit that you used for sketch1401 in Hour 14, "Working with Digital Interfaces." First, build the hardware for the example:

1. Keep the potentiometer connected to analog interface A0 that you used for the sketch1501 circuit on the breadboard.

2. Add the traffic signal circuit that you used from sketch1401. The final circuit diagram is shown in Figure 15.4. You can use three different-colored LEDs, or you can use three of the same color LED.

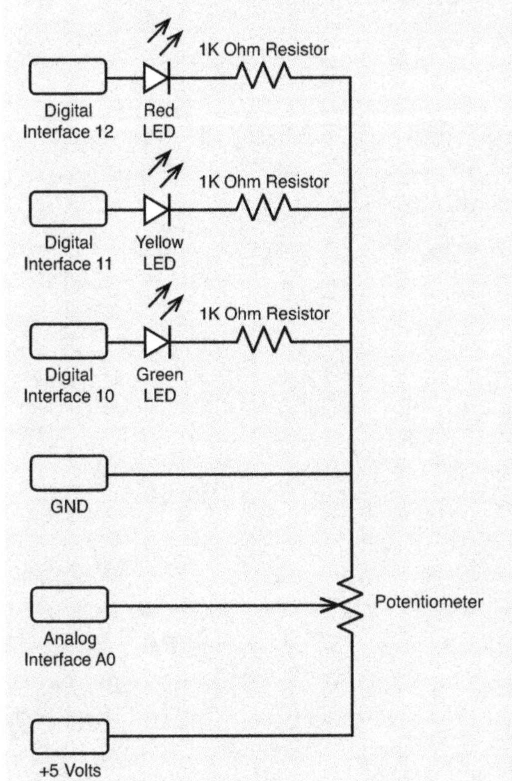

FIGURE 15.4
The complete traffic signal diagram.

The three LED's are connected to digital interfaces 10, 11, and 12. You control each LED using the `digitalWrite` function to provide a `HIGH` signal to light the LED, or a `LOW` signal to turn off the LED. The next step is to build the sketch to control the LEDs using the potentiometer. Follow these steps for that:

1. Open the Arduino IDE, and then enter this code into the editor window:

```
int input;
void setup() {
   pinMode(10, OUTPUT);
   pinMode(11, OUTPUT);
   pinMode(12, OUTPUT);
   digitalWrite(10, LOW);
   digitalWrite(11, LOW);
   digitalWrite(12, LOW);
}

void loop() {
   input = map(analogRead(A0), 0, 1023, 1, 3);
   switch(input) {
     case 1:
       stoplight();
       break;
     case 2:
      yieldlight();
      break;
     case 3:
      golight();
      break;
   }
}

void stoplight() {
   digitalWrite(10, LOW);
   digitalWrite(11, LOW);
   digitalWrite(12, HIGH);
}

void yieldlight() {
   digitalWrite(10, LOW);
   digitalWrite(11, HIGH);
   digitalWrite(12, LOW);
}

void golight() {
   digitalWrite(10, HIGH);
   digitalWrite(11, LOW);
   digitalWrite(12, LOW);
}
```

2. Save the sketch code as **sketch1502**.
3. Click the Upload icon to verify, compile, and upload the sketch to your Arduino unit.
4. Turn the potentiometer control arm to different positions and watch the LEDs as they light up.

As you turn the potentiometer control arm, the `analogRead` function returns a different value from 0 to 1023. The `map` function maps that result to the 1 to 3 range. The `switch` function executes the appropriate `case` statement based on the mapped value. Each `case` statement runs a separate function to light one of the three LEDs.

Changing the Reference Voltage

By default, the Arduino assigns the largest digital value (1023) when a +5 voltage value is present on the analog input. This is called the reference voltage. The digital output values are based on the percentage of the reference voltage the analog input signal is.

The Arduino allows you to change the reference voltage to use a different scale. Instead of 1023 being assigned to +5 volts, you can change it so the `analogRead` function produces the 1023 value at +1.1 volts. You do that using the `analogReference` function:

```
analogReference(source);
```

The *`source`* parameter value specifies what reference to use for the analog-to-digital conversion. You can use either a built-in internal reference source for the Arduino, or you can use your own external reference source. This section shows both methods for changing the reference voltage used on the Arduino.

Using an Internal Reference Voltage

The Arduino can provide a separate internal reference voltage for analog input signals besides the default +5 volts. To do that, you specify the `INTERNAL` label for the `analogReference` function:

```
analogReference(INTERNAL);
```

For most Arduino models, the internal reference voltage is +1.1 volts. The Arduino Mega model uses a separate internal reference voltage that can generate two separate values: +1.1 or +2.56 volts. Because the Mega can generate two separate input reference voltages, it has two separate labels:

- `INTERNAL1V1` for the +1.1 volt reference
- `INTERNAL2V56` for the +2.56 volt reference

When you change the reference voltage, the Arduino bases the 0 through 1023 digital value on the new reference voltage. So, when the input voltage is 1.1, it returns a 1023 value.

Using an External Reference Voltage

You can also provide your own reference voltage for the `analogRead` function. You do that by using the `AREF` pin on the Arduino header. Just apply the voltage that you want to use as the maximum value to the `AREF` pin, and set the `analogReference` function to `EXTERNAL`:

```
analogReference(EXTERNAL);
```

Now when the input analog voltage reaches the reference voltage value, the Arduino returns a 1023 value, and scales the other voltages accordingly.

WATCH OUT!

Reference Voltage Values

The Arduino can only handle input values of +5 volts or less, so you can't increase the reference voltage to more than +5 volts. If you need to detect voltages larger than +5 volts, you must use a resistor to decrease the input voltage to within the acceptable input range for the Arduino. Hour 18, "Using Sensors," dives into this topic in more detail.

Analog Output

The PWM feature on the Arduino enables you to simulate an analog output voltage on specific digital interfaces. The function to use the PWM feature is the `analogWrite`:

```
analogWrite(pin, dutycycle);
```

The *`pin`* parameter specifies the digital interface number to use. You don't need to use the `pinMode` function to set the digital interface mode; the `analogWrite` function will do that for you automatically.

The *`dutycycle`* parameter specifies the amount of time the digital pulse is set to `HIGH`. What complicates things is that the duty cycle specified for the `analogWrite` function isn't a percentage, but a value from 0 to 255. The 255 value creates a 100% duty cycle signal, generating a +5 analog voltage from the digital interface. The values between 0 and 255 scale the analog output voltage accordingly.

Using the Analog Output

The following example demonstrates how to use the PWM ports on an Arduino to output an analog signal.

TRY IT YOURSELF ▼

Generating an Analog Output

In this example, you use a potentiometer to control the brightness of two LEDs connected to PWM interfaces on the Arduino. You use the same circuit that you built for the sketch1502 example. As it turns out, digital interfaces 10 and 11 that we used for two of the traffic signal interfaces also support PWM output. That means you can control the voltage applied to them using the `analogWrite` function.

You just need to write a sketch that reads the analog value from the potentiometer, and then changes the output voltages on digital interfaces 10 and 11 based on that analog voltage. Follow these steps to do that:

1. Open the Arduino IDE, and then enter this code into the editor window:

```
int input;
void setup() {
}

void loop() {
   input = map(analogRead(A0), 0, 1023, 0, 255);
   analogWrite(10, input);
   analogWrite(11, (255-input));
}
```

2. Save the sketch code as **sketch1503**.
3. Click the Upload icon to verify, compile, and upload the sketch code to your Arduino unit.
4. Rotate the potentiometer control arm to change the voltage value present on the analog input pin.

The sketch1503 code maps the input value received from the A0 analog input to the 0 through 255 range, and then uses that value as the duty cycle for the PWM signal send out on digital interfaces 10 and 11. Notice that the duty cycle for interface 11 will be the opposite as that applied to interface 10. (When the `input` value is 0, interface 11 will have a duty cycle of 255.) This will cause one LED to get brighter as the other gets dimmer, and vice versa.

As you rotate the potentiometer control arm, you should notice that the LEDs get brighter or dimmer. At one end of the rotation, one LED should go completely out, showing 0 volts on the analog output, and the other should go to maximum brightness, showing +5 volts on the analog output. Then at the opposite end of the rotation, the opposite lights should go out and get bright.

Summary

This hour explored how you can use the Arduino to read and generate analog signals. The Arduino Uno supports six analog input interfaces, labeled A0 through A5. You use the analogRead function to read values from those interfaces. The `analogRead` function returns an integer value from 0 to 1023, based on the signal level present on the interface. You can change the range of the values using either the `constrain` or `map` functions.

All Arduino models support generating analog output signals using pulse-width modulation. You use the `analogWrite` function to do that. The `analogWrite` function specifies the digital interface to use for the output, along with the duty cycle, which determines the level of the analog output signal.

This next hour shows you how to use interrupts in your Arduino programs. Interrupts enable you to interrupt your sketch whenever a signal change is detected on an interface. Using interrupts, you don't have to worry about constantly polling an interface to wait for a signal change.

Workshop

Quiz

1. What function should you use to change the range of the `analogRead` output?

   A. `pinMode`

   B. `map`

   C. `switch`

   D. `case`

2. The `analogWrite` function uses a DAC on the Arduino Uno to convert the digital value to an analog value. True or false?

3. What code would you write to remap the output of analog interface A2 to a range of 1 to 10?

Answers

1. B. The map function allows you to map the original range of 0 to 1023 from the `analogRead` output to another range.

2. False. The `analogWrite` function uses pulse-width modulation to simulate an analog output signal using the digital interfaces.

3. You could use the following line:

```
map(analogRead(A2), 0, 1023, 1, 10);
```

Q&A

Q. Can the Arduino output larger analog voltages?

A. No, the Arduino can only output +5 volts. If you need to generate larger analog voltages, you must create an external circuit that can increase the analog output generated by the Arduino.

Q. How much current does the Arduino generate for the output analog signal?

A. The Arduino generates 40mA of current from the PWM signal.

HOUR 16
Adding Interrupts

What You'll Learn in This Hour:

- What are interrupts
- How to use external interrupts
- How to use pin change interrupts
- How to use timer interrupts

Hour 14, "Working with Digital Interfaces," showed you the basics of how to read data from a digital interface. Sometimes how your sketch reads data on the digital interface can be important. You might sometimes need to detect exactly when a specific action happens on an external circuit, such as when a sensor signal transitions from a `LOW` to a `HIGH` value. If your sketch just checks the interface periodically, it may miss the transition when it happens. Fortunately, the Arduino developers have created a way for our sketches to watch for events on digital interfaces, and handle them when they occur. This hour discusses how to use interrupts to watch for events and demonstrates how to use them in your sketches.

What Are Interrupts?

So far in your sketches, you've been writing code that checks the digital interface at a specific time in the sketch. For example, in the traffic signal sketch in Hour 14, the code only checks the button position after the yellow LED lights. This method of retrieving data from an interface is called polling.

With polling, the sketch checks, or polls, a specified interface at a specific time to see if a signal has changed values. The more times you poll the interface in your sketch loop, the quicker the sketch can respond to a signal change on the interface.

However, a limit applies to that. Your sketch can't poll the interface all the time; it might have to do other things in the meantime.

A more practical way of watching for interface signal changes is to let the interface itself tell you when the signal changes. Instead of having the sketch constantly poll the interface looking for a signal change, you can program the sketch to interrupt whatever it's doing when the signal changes on the interface. This is aptly called an interrupt.

Interrupts trigger actions based on different types of changes in the digital signal value. You can program your sketch to watch for three basic types of signal changes:

- Falling signal values from `HIGH` to `LOW`
- Rising signal values from `LOW` to `HIGH`
- Any type of signal change from one value to the opposite value

When the Arduino detects an interrupt, it passes control of the sketch to a special function, called an interrupt service routine (ISR). You can code the ISR to perform whatever actions need to happen when the interrupt triggers.

More than one type of interrupt is available on the Arduino. The next section describes the different types and discusses when to use each type.

Types of Interrupts

The ATmega microprocessor chip used in the Arduino supports three types of interrupts:

- External interrupts
- Pin change interrupts
- Timer interrupts

Each type of interrupt works the same way, but is generated by a different trigger in the Arduino. This section discusses the differences between these three types of interrupts.

External Interrupts

External interrupts are triggered by a change in a digital interface signal connected to the Arduino. The ATmega microprocessor chip has built-in hardware support for handling external interrupts. The hardware continually monitors the signal value present on the digital interface and triggers the appropriate external interrupt when it detects a signal change.

By using a hardware interface, external interrupts are very fast. As soon as the digital interface detects the signal change, the microprocessor triggers the external interrupt that your sketch can catch.

The hardware external interrupts are referenced separate from the digital interfaces that they monitor. The external interrupts are named `INTx`, where *x* is the interrupt number. The first external interrupt is `INT0`. Those are the numbers you'll need to use in your sketches to reference the external interrupt to monitor.

The downside to external interrupts is that not all digital interfaces have the hardware to generate them. Each Arduino model has a limited number of digital interfaces that support generating external interrupts. Table 16.1 shows the digital interface support for external interrupts on the different Arduino models.

TABLE 16.1 Arduino Support for External Interrupts

Model	INT0	INT1	INT2	INT3	INT4	INT5
Uno	2	3	—	—	—	—
Leonardo	3	2	0	1	—	—
Mega	2	3	21	20	19	18

For example, on the Arduino Uno model, the `INT0` interrupt signal monitors digital interface 2, and the `INT1` interrupt signal monitors digital interface 3. Those are the only two digital interfaces on the Uno model that support external interrupts. The Leonardo Arduino model can support four external interrupts, and the Mega Arduino model can support six.

Because external interrupts are generated by the ATmega microprocessor, support for them is already built in to the Arduino programming language. You don't need to load any external libraries to use external interrupts in your sketches.

Pin Change Interrupts

Pin change interrupts are software driven instead of hardware driven. The benefit of pin change interrupts is that the Arduino generates a pin change interrupt for any type of signal change on any interface. Although this enables you to detect signal changes from any interrupt, there's a catch.

The catch is that your sketch code must determine just why the pin change interrupt was generated. It has to decode which interface generated the pin change interrupt, and based on what type of signal change.

Fortunately for us, some Arduino developers have created the PinChangeInt library to help out with decoding pin change interrupts that the Arduino generates. It allows you to monitor specific interrupts and to specify the ISR the Arduino runs when the interrupt occurs. This makes handling pin change interrupts almost as easy as working with external interrupts.

Unfortunately, the Arduino IDE environment doesn't install the PinChangeInt library by default, so you'll have to do some work to get that installed.

Timer Interrupts

The last type of interrupt that you can use in the Arduino is timer interrupts. Instead of triggering an interrupt based on an event on a digital interface, timer interrupts trigger an interrupt based on a timing event, sort of like an egg timer.

With timer interrupts, your sketch can define an ISR function to trigger at a preset time in the sketch code. When the timer in the Arduino reaches that time, the Arduino automatically triggers the ISR function and runs the function code.

Timer interrupts can come in handy if you need to perform tasks at preset times in the sketch, such as recording a sensor value every 5 minutes. You just set a timer interrupt to trigger every 5 minutes, and then set code in the ISR function to read and record the sensor value.

Using External Interrupts

Coding external interrupts in your sketches is a fairly straightforward process. To use external interrupts, you just need to add two things to your sketch:

- Set the external interrupt to watch
- The ISR function to run when the interrupt triggers

The first step is to define which external interrupt to watch. You do that using the `attachInterrupt` function:

```
attachInterrupt(interrupt, isr, mode);
```

The *`interrupt`* parameter defines the external interrupt signal to monitor. For the Arduino Uno model, this value will be either 0 for monitoring interrupt `INT0` on digital interface 2, or 1 for monitoring interrupt `INT1` digital interface 3.

The *`isr`* parameter defines the name of the ISR function the Arduino runs when the external interrupt triggers. The Arduino halts operation in the main sketch code and immediately jumps to the ISR function. When the ISR function finishes, the Arduino jumps back to where it left off in the main sketch code.

The *`mode`* parameter defines the signal change that will trigger the interrupt. You can set four types of external interrupt modes:

- **`RISING`**—Signal changes from `LOW` to `HIGH`.
- **`FALLING`**—Signal changes from `HIGH` to `LOW`.

- **CHANGE**—Any type of signal value change.
- **LOW**—Triggers on a LOW signal value.

The LOW interrupt trigger can be somewhat dangerous because it will continually trigger as long as the interface signal is at a LOW value, so be careful with that one. The others are fairly self-explanatory.

WATCH OUT!

ISR Speed

You want to be aware of a few do's and don'ts regarding the code that you can use in an ISR function. The idea is to make the ISR code as short and fast as possible, because the Arduino holds up all other processing while running the ISR function. While running the ISR function, the Arduino also ignores any other interrupts that may be generated, so your sketch could miss other interrupts in the process.

Also, don't use the Serial library methods to output text to the serial monitor inside of an ISR function. The Serial library uses interrupts to send data out the serial ports, which won't work in the ISR function because the Arduino ignores interrupts inside the ISR function!

When your sketch is finished monitoring external interrupts, you should use the detach Interrupt function to disable them:

```
detachInterrupt(interrupt);
```

Just specify the interrupt number that you previously set in the attachInterrupt function to disable it.

Testing External Interrupts

Now that you've seen the basics on how to use external interrupts, let's work on an example that demonstrates using them in an Arduino sketch.

TRY IT YOURSELF ▼

Using External Interrupts in a Sketch

In this example, you revisit the traffic signal sketch that you created in Hour 14 in the "Experimenting with Digital Input" section. If you remember, sketch1402 used a switch to change the time value assigned to the green LED state. The downside was that you had to press and hold the switch at a specific time in the sketch for it to detect the change.

For this example, you use an external interrupt to control the LED timing. This allows you to press the switch at any time in the traffic light cycle to change the green LED time allocation.

This example requires that you rebuild the traffic signal circuit used for sketch1402 from Hour 14, but moving the switch to the digital interface 2 pin on the Arduino to trigger the external interrupt. The components required for the example are as follows:

- Three LEDs (one red, one yellow, and one green)
- Four 1K ohm resistors (color code brown, black, red)
- A momentary contact switch
- A breadboard
- Six jumper wires

First, follow these instructions for building the traffic signal hardware required for the example:

1. Place the three LEDs on the breadboard so that the short leads are all on the same side and that the two leads straddle the space in the middle of the board so that they're not connected. Place them so that the red LED is at the top, the yellow LED in the middle, and the green LED is at the bottom of the row.
2. Connect a 1K ohm resistor between the short lead on each LED to a common rail area on the breadboard.
3. Connect a jumper wire from the common rail area on the breadboard to the GND interface on the Arduino.
4. Connect a jumper wire from the green LED long lead to digital interface 10 on the Arduino.
5. Connect a jumper wire from the yellow LED long lead to digital interface 11 on the Arduino.
6. Connect a jumper wire from the red LED long lead to digital interface 12 on the Arduino.
7. Place the momentary contact switch on the breadboard under the LEDs so that the switch straddles the center divide of the breadboard.
8. Connect a jumper wire from digital interface 2 on the Arduino to one of the switch pins.
9. Connect the other switch pin to the GND rail on the breadboard.
10. Connect a jumper wire from the 5V pin on the Arduino to a common rail area on the breadboard.
11. Connect a 1K ohm resistor from the row that has the left side of the switch pin connected to digital interface 2 to the breadboard 5V common rail area.

BY THE WAY

Resistors and LEDs

The size of the resistor to use with the LEDs depends on the amount of current the LEDs you use can handle. The Arduino outputs 40mA of current, so for most larger electronic circuit LEDs, you need only a 1K ohm resistor to limit the current. If you use a smaller-sized LED such as the ones built in to the Arduino, you must use a larger-sized resistor, such as a 10K ohm, to limit the current applied to the LED.

That completes the electronics for the circuit. Figure 16.1 shows the circuit diagram for the completed circuit.

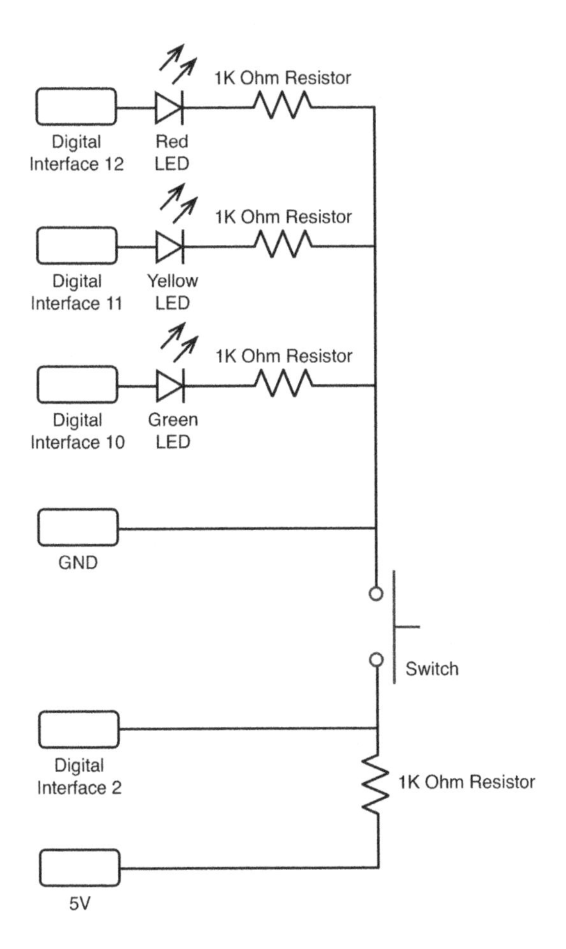

FIGURE 16.1
The external interrupt test circuit.

This circuit re-creates our standard traffic signal light setup using digital interfaces 10, 11, and 12 to control the traffic signal LEDs, plus adds a switch to digital interface 2 to trigger the `INT0` interrupt.

The switch uses a 1K ohm resistor as a pullup resistor to tie digital interface 2 to a `HIGH` signal when the switch isn't pressed. When you press the switch, it pulls the interface `LOW`.

The circuit uses the built-in LED on digital interface 13 to monitor when the interrupt is triggered. (Remember, you can't use the `Serial.print` function inside an ISR function.) When the L led on the Arduino unit lights, the timer is set to a 10-second interval for the green LED.

Now you're ready to create the sketch for the example. Follow these steps to do that:

1. Open the Arduino IDE, and then enter this code into the editor window:

```
int stop = 6;
int yield = 2;
int go = 6;
void setup() {
   Serial.begin(9600);
   pinMode(13, OUTPUT);
   pinMode(10, OUTPUT);
   pinMode(11, OUTPUT);
   pinMode(12, OUTPUT);
   digitalWrite(8, LOW);
   digitalWrite(10, LOW);
   digitalWrite(11, LOW);
   digitalWrite(12, LOW);
   attachInterrupt(0, changeTime, FALLING);
}

void loop() {
   stoplight(stop);
   golight(go);
   yieldlight(yield);
}

void stoplight(int time) {
   digitalWrite(10, LOW);
   digitalWrite(11, LOW);
   digitalWrite(12, HIGH);
   Serial.println("Light mode: Stop");
   delay(time * 1000);
}

void yieldlight(int time) {
   digitalWrite(10, LOW);
   digitalWrite(11, HIGH);
```

```
    digitalWrite(12, LOW);
    Serial.println("Light mode: Yield");
    delay(time * 1000);
}

void golight(int time) {
    digitalWrite(10, HIGH);
    digitalWrite(11, LOW);
    digitalWrite(12, LOW);
    Serial.print("Light mode: Go - ");
    Serial.println(time);
    delay(time * 1000);
}

void changeTime() {
    if (go == 6) {
      go = 10;
      digitalWrite(13, HIGH);
    } else {
      go = 6;
      digitalWrite(13, LOW);
    }
}
```

2. Save the sketch as **sketch1601**.
3. Click the Upload icon to verify, compile, and upload the sketch to your Arduino unit.
4. Open the serial monitor to run the sketch and view the output.

You should recognize most of the code in this sketch from Hour 14. It creates separate functions to control the stop, yield, and go states of the traffic signal, just like before. However, this version uses the `attachInterrupt` function to tell the Arduino to monitor the input on digital interface pin 2 (`INT0`). Digital interface 2 is wired in a pullup circuit, providing a `HIGH` signal value by default. When you press the switch button, the circuit pulls the digital interface 2 pin to ground, causing the signal to fall to a `LOW` value, triggering the `FALLING` interrupt on `INT0`.

The `attachInterrupt` function catches the falling interrupt, and runs the `changeTime` ISR function defined in the sketch. The `changeTime` function changes the value of the `go` variable, which controls how long the green LED stays lit. It also changes the fourth LED to indicate when the `go` variable is 10 (the LED will light) or 6 (the LED will be off). This feature allows you to peek inside the ISR function and easily see when it triggers.

When you run the sketch, you can press the switch button at any time in the traffic light cycle to change the length assigned to the green LED.

WATCH OUT!

Switch Bounce

One downside to using interrupts is that they are very sensitive. Every time the interface signal changes, it generates a new interrupt. You may have noticed that if you're not careful pressing the switch button, it may trigger two or more interrupts with one press. This is called switch bounce.

To help minimize switch bounce, place a capacitor across the two switch pins. The capacitor stores electricity when the switch is open, and discharges when the switch is closed. The change won't actually register on the digital interface until the capacitor fully discharges.

If the switch bounces while the capacitor is discharging, the discharge will override the bounce, forcing the signal to stay `HIGH`. This helps mask short switch bounces in the circuit, but it does reduce the sensitivity of the switch.

Using Pin Change Interrupts

Thanks to the PinChangeInt library, using pin change interrupts isn't too different from using external interrupts. You just specify the pin number that you want to monitor, the type of signal change you want to detect, and the ISR function to run when the interrupts occurs.

This section shows you how to install the PinChangeInt library and use it in your sketches.

Installing the PinChangeInt Library

The PinChangeInt library is packaged a little differently than the standard Arduino libraries, making it a little harder to import into your Arduino IDE environment. Here are the steps you need to follow to import the PinChangeInt library into your Arduino IDE environment:

1. Open a browser and navigate to the PinChangeInt web page: http://code.google.com/p/arduino-pinchangeint/.
2. Click the Downloads tab at the top of the main web page.
3. From the Downloads page, click the Download icon for the latest nonbeta version available (at the time of this writing, version 1.72).
4. Click the Save button to download the zip file to your workstation.
5. Open the zip file on your system, and extract the PinChangeInt folder from the zip file to a temporary location.
6. Move the PinChangeInt folder from the temporary location to the libraries folder under your Arduino installation location (normally under the Documents\Arduino folder for your user account in both Windows and OS X).
7. If you have the Arduino IDE open, close it and reopen it so it can recognize the new PinChangeInt library.

WATCH OUT!

Using the Import Library

Unfortunately, the PinChangeInt zip file contains three separate library folders, so you can't use the Arduino IDE Import Library feature shown in Hour 13, "Using Libraries." Instead, you have to extract the PinChangeInt folder from the distribution zip file and manually copy it into the libraries folder for the Arduino IDE.

The PinChangeInt library is now imported into your Arduino IDE environment. When you go to the Import Library submenu, the PinChangeInt library should appear under the Contributed section of the library listing.

Experimenting with Pin Change Interrupts

After importing the PinChangeInt library into your Arduino IDE environment, you can easily use it to track interrupts on any digital interface on the Arduino. First, you must import the library into your sketch by using the Import Library option of the Sketch menu item. This adds a `#include` directive to your sketch:

```
#include <PinChangeInt.h>
```

Next, you need to set the digital interface that you want to use for the interrupt trigger to input mode:

```
pinMode(7, INPUT);
```

Finally, you use the special `PCintPort` object that's created in the library to access the `attachInterrupt` method. That line looks like this:

```
PCintPort::attachInterrupt(7, changeTime, FALLING);
```

With the PinChangeInt `attachInterrupt` method, you specify the pin number of the digital interface as the first parameter, the ISR function to call as the second parameter, and the signal change to monitor as the third parameter.

Let's go through an example of using the PinChangeInt library in your sketch.

TRY IT YOURSELF ▼

Using the Pin Change Interrupt

You can easily convert the external interrupt example used in sketch1601 to use a pin change interrupt. Use the circuit that you built for the sketch1601 example, and follow these steps:

1. Open the Arduino IDE, and then open the sketch1601 sketch.
2. Select Sketch from the menu bar at the top of the Arduino IDE.

3. Select Import Library, and then select the PinChangeInt library.
4. Change the `setup` function code to look like this:

```
void setup() {
   Serial.begin(9600);
   pinMode(7, INPUT);
   pinMode(13, OUTPUT);
   pinMode(10, OUTPUT);
   pinMode(11, OUTPUT);
   pinMode(12, OUTPUT);
   digitalWrite(13, LOW);
   digitalWrite(10, LOW);
   digitalWrite(11, LOW);
   digitalWrite(12, LOW);
   PCintPort::attachInterrupt(7, changeTime, FALLING);
}
```

5. Save the new sketch as **sketch1602**.
6. Move the jumper wire from the switch that was connected to digital interface 2 to digital interface 7.
7. Click the Upload icon to verify, compile, and upload the sketch to your Arduino unit.
8. Open the serial monitor to run the sketch and view the output.

The updated code sets a pin change interrupt on digital interface 7 and watches for a falling signal change. When you press the switch, it triggers the pin change interrupt, and changes the time allocated to the green LED in the traffic signal. Now with pin change interrupts, you can place the switch on any digital interrupt and monitor when it changes.

Working with Timer Interrupts

Timer interrupts allow you to set interrupts that trigger at a preset time period while your sketch runs. All Arduino models include at least one timer that you can access with your sketch code, called Timer1. The Arduino Mega model provides three separate timers that you can use.

To access the timer, you need to use special software. Again, the Arduino developer community has come to our rescue and has provided an easy-to-use library for us. The Timer One library provides everything we need to utilize the Timer1 internal timer in your Arduino unit.

This section demonstrates how to get and install the Timer One library, plus shows an example of how to use it in your own Arduino sketches.

Downloading and Installing Timer One

Similar to the PinChangeInt library, you'll have to go out to the Internet and download the Timer One library to install it in your Arduino IDE environment. Here are the steps to do that:

1. Open a browser and navigate to the Timer One Web page: http://code.google.com/p/arduino-timerone/.
2. Click the Downloads link at the top of the web page.
3. Click the Download icon for the latest release version of the Timer One library (called TimerOne-v9.zip at the time of this writing).
4. Save the file to a location on your workstation.
5. Open the downloaded zip file, and then extract the TimerOne-v9 folder to a folder called TimerOne in a temporary location (you'll have to rename the folder to remove the dash for it to work).
6. Copy the TimerOne folder to the libraries folder in your Arduino IDE environment.

After importing the Timer One library into your Arduino IDE, you're ready to use it in your sketches.

Testing Timer Interrupts

Let's work through an example of using the Timer One library. For this example, you create a sketch that blinks the built-in LED connected to the digital interface 13 pin on the Arduino using the Timer One library. Just follow these steps:

1. Open the Arduino IDE, click the Sketch menu bar item, and then under Import Library, select the Timer One library.
2. Enter this code into the editor window:

```
#include <TimerOne.h>

int state = 0;
void setup() {
  pinMode(13, OUTPUT);
  digitalWrite(13, state);
  Timer1.initialize(1000000);
  Timer1.attachInterrupt(blinkme);
}

void loop() {
}
```

```
void blinkme() {
  state = !state;
  digitalWrite(13, state);
}
```

3. Save the sketch as **sketch1603**.

4. Click the Upload icon to verify, compile, and upload the sketch to your Arduino unit.

The sketch1603 code uses methods in the `Timer1` class object defined in the Timer One library. This sketch uses two methods from that library:

```
Timer1.initialize(time);
Timer1.attachInterrupt(isr);
```

The `initialize` method defines the amount of time between triggering the timer interrupt. The *`time`* value is defined in microseconds, so to trigger the interrupt every second, you'll need to enter the value 1000000.

The `attachInterrupt` method defines the ISR function to call when the timer interrupt triggers.

In this example, I created an integer variable called `state` that toggles between `HIGH` and `LOW` values on each timer interrupt. When the `state` variable is set to a `HIGH` value, the LED that is connected to digital interface 13 lights up. When the `state` variable is set to a `LOW` value, the LED is not lit. The timer interrupt triggers the `blinkme` function every 1 second, which causes the LED to change states every second. Feel free to change the value set for the timer interval and watch how it changes the blink rate of the LED.

Ignoring Interrupts

Sometimes you might not want your sketch code interrupted by events (such as a time-sensitive part of the sketch code). In those situations, you can disable interrupts so that they don't trigger the ISR function assigned to them. You do that using the `nointerrupts` function:

```
nointerrupts();
```

As you can see, there aren't any parameters to the `nointerrupts` function, so you can't disable a specific interrupt. When you run the `nointerrupts` statement, all interrupts on the Arduino are ignored.

You can reenable interrupts when your sketch is ready to accept them again using the `interrupts` function:

```
interrupts();
```

Again, as you can see, the `interrupts` function doesn't use any parameters, so it enables all interrupts on the Arduino.

Summary

This hour explored the world of interrupts on the Arduino. Interrupts alter the operation of a sketch when a specified event is detected on the Arduino. You can use three types of interrupts on the Arduino. External interrupts are built in to the ATmega microprocessor chip and trigger an interrupt when the microprocessor detects a signal change on a specific digital interface. Most Arduino models support only two external interrupts.

Pin change interrupts are software generated by the microprocessor, but can be set for any digital interface. Unfortunately the sketch must decode the cause of the pin change interrupt, which can be tricky. The PinChangeInt library provides simple functions for us to use to interface with pin change interrupts.

Timer interrupts allow you to set a timer to trigger an interrupt at predefined intervals in the sketch. All Arduino models support one timer interrupt, and the Due, Mega, and Yun models support multiple interrupts.

The next hour covers how to use serial communication techniques to help your Arduino communicate with the outside world. The Arduino supports three different types of serial communication protocols, as discussed in the next hour.

Workshop

Quiz

1. Which type of interrupt should you use to read a sensor value every 5 minutes in your sketch?

    A. External interrupt

    B. Pin change interrupt

    C. Timer interrupt

    D. Polling

2. You can use external interrupts on any digital interface on the Arduino. True or false?

3. If you set both the `INT0` and `INT1` external interrupts, how can you turn off only the `INT1` interrupt and keep the `INT0` interrupt active?

Answers

1. C. Timer interrupts allow us to set a predefined interval for the interrupt to trigger an ISR function. You can then place code to read the sensor value inside the ISR function.
2. False. Most Arduino models only support two external interrupts, on digital pins 2 and 3.
3. You can use the `detachInterrupt(1)` function to stop receiving interrupts for `INT1`, but continue receiving interrupts for `INT0`.

Q&A

Q. **Can you define different ISR functions for different interrupts, or must all the interrupts use the same ISR function?**

A. You can define multiple ISR functions and assign each one to a different interrupt.

Q. **Can you change the mode of an interrupt from `RISING` to `FALLING`?**

A. No, you must detach the interrupt first, and then attach a new interrupt with the new mode.

Q. **Can I use both external and pin change interrupts at the same time?**

A. Yes, but be careful that you don't assign both to the same digital interface.

HOUR 17

Communicating with Devices

What You'll Learn in This Hour:

- The different types of serial communication protocols
- Using the serial interface pins on the Arduino
- How the SPI interface works
- How to talk to another device using I^2C

More likely than not, you'll run into a situation where you want your Arduino to communicate with some type of external device. Whether you're interfacing your Arduino with the serial monitor tool in the Arduino IDE or passing data to an LCD controller to display information, communication is a vital part of working with the Arduino. This hour takes a look at three common methods of communicating with an Arduino.

Serial Communication Protocols

The basic mode of communication among Arduino devices is serial communication. Serial communication requires a fewer number of connections between the devices than other communication protocols (often with just two or three lines), and provides a simple way to move data between disparate devices.

The Arduino hardware design includes three different types of serial communication methods:

- A standard serial port
- A Serial Peripheral Interface (SPI) port
- An Inter-Integrated Circuit (I^2C) port

These are all common serial communication methods used in the electronics industry. By providing all three methods, the Arduino developers have made it easy to connect your Arduino with just about any type of external device you'll run into. The Arduino IDE software even includes libraries that make it easy for you to interface with each of these different communication methods in your sketches.

The following sections examine each of the different serial communication methods, showing you how to use each of them in your sketches.

Using the Serial Port

Each Arduino model contains at least one standard serial port interface. The serial port interface sends data as a serial sequence of bits to the remote device. Because data is sent one bit at a time, the serial port only needs two pins to communicate—one to send data to the remote device, and one to receive data from the remote device.

This is the interface that the Arduino IDE software uses to transfer data into your Arduino unit, as well as receive simple output from your Arduino unit. You can easily adapt the standard serial interface to communicate with other serial devices as well.

The Serial Port Interfaces

The serial interface on the Arduino uses two digital interface pins for serial communication with external devices. By default, all Arduino models use digital interface pins 0 and 1 to support the primary serial interface. The Arduino uses pin 0 as the receive port (called RX), and pin 1 for the transmit port (called TX).

The Arduino software names this port serial, which is also the name of the object you use in your sketches to send and receive data using the port (such as when you used the `Serial.print` function in your sketches).

Because the serial interface is commonly used to transfer your sketch code into the Arduino, to make life even easier for us the Arduino developers connected the serial interface pins to a serial-to-USB adapter built in to the Arduino unit. This is what allows you to plug your Arduino directly into to your workstation using a standard USB cable.

WATCH OUT!

Using the USB Serial Interface

Because the USB serial interface uses digital interfaces 0 and 1, you can't use those interfaces in your sketches as digital inputs or outputs once you use the `Serial.begin` function to initialize the serial interface to output data to the serial monitor. You can use the `Serial.end` function to stop the serial interface and return the interfaces back to their normal functions.

The Arduino Due and Mega models also include three other serial port interfaces. Digital interface pins 18 and 19 are used for the Serial1 port, pins 16 and 17 for Serial2, and pins 14 and 15 are used for Serial3. These serial ports don't have a serial-to-USB adapter connected to them, so you'll either need to provide your own, or just use the raw serial pins in your circuits.

Because the serial ports use two separate pins to communicate, you have to be careful how you connect the Arduino serial port to other devices. If you connect your Arduino to another Arduino unit, or another type of serial device using the serial port, remember to cross-connect the interface pins. The receive pin on the Arduino must connect to the transmit port on the external device, and vice versa, as shown in Figure 17.1.

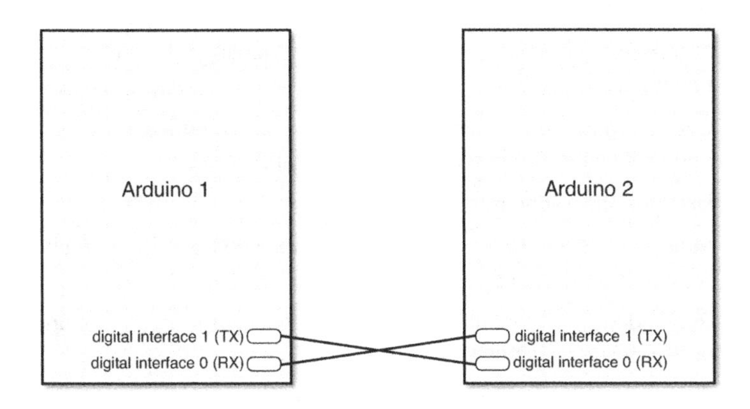

FIGURE 17.1
Connecting two Arduino units using the serial interface pins.

The Arduino sends and receives serial signals using Transistor-transistor-logic (TTL)-level voltages. This means that the 1 bit is represented by +5 volts, and a 0 bit is represented by 0 volts. Make sure that the remote device you connect to your Arduino serial interface also supports TTL-level signaling.

WATCH OUT!

RS-232 Serial Interfaces

Don't try to connect your Arduino serial port to a standard RS-232 serial interface, such as what is found in the COM ports on older desktop workstations. The RS-232 interface standard uses +12V for the signal, which will damage the digital interfaces on your Arduino!

The Serial Library Functions

The Serial library provides a set of functions for you to easily send and receive data across the serial interfaces on your Arduino. You've already seen them in action as we used the serial interface to send data to the serial monitor utility in the Arduino IDE.

Not only can you send data out the serial interface, but the Serial library also includes functions that enable you to read data received on the serial interface. This allows you to send data from the serial monitor back into your Arduino sketch to control things. This is a great way to push data into your sketch as it runs.

Table 17.1 shows the functions provided by the Serial library that you can use in your sketches.

TABLE 17.1 The Serial Functions

Function	Description
`available()`	Returns the number of bytes available for reading from the serial port.
`begin(rate[, config])`	Sets the data rate (in bits per second) for the interface, and optionally sets the protocol data, parity, and stop bits.
`end()`	Disables the serial interface.
`find(string)`	Reads data from the interface until the specified *string* is found. Returns true if the string is found.
`findUntil(string, terminal)`	Reads data from the interface until the specified *string* is found, or the *terminal* string is found.
`flush()`	Waits until all data is sent out of the interface.
`parseFloat()`	Returns the first valid floating-point value from the interface.
`parseInt()`	Returns the first valid integer value from the interface.
`peek()`	Returns the next byte of data from the interface, but doesn't remove it from the interface buffer.
`print(text)`	Sends data to the interface as ASCII text.
`println(text)`	Sends data to the interface as ASCII text, and terminates it with carriage return and newline characters.
`read()`	Returns the first byte of incoming data from the interface, or –1 if no data is present.
`readBytes(buffer, length)`	Returns *length* bytes of incoming data into the *buffer* array, or 0 if no data is present.
`readBytesUntil(char, buffer, length)`	Returns *length* bytes of data into the *buffer* array until the *char* character is detected.
`setTimeout(time)`	Sets the number of milliseconds to wait for data when using the *readBytes* or *readBytesUntil* functions. The default is 1000 milliseconds.
`write(val)`	Sends the *val* string as a series of bytes.

You've already seen some of these functions in action as we worked on the experiment sketches in the previous hours. The following sections go through a brief rundown of the more commonly used `Serial` functions that you'll want to use in your sketches.

Starting Communications

As you've already seen in our experiment sketches, to start communicating using the serial interface, you must use the `Serial.begin` function. This function initializes the digital interface pins for serial mode, and sets the communication parameters that the serial interface uses:

```
Serial.begin(rate[, config]);
```

The first parameter is required; it sets the speed of the data transfer in bits per second (called the baud rate). The Arduino serial interface supports baud rates up to 115,200 bits per second, but be careful, because using higher baud rates can sometimes introduce errors, especially in longer connection wires. It's common practice to use 9600 baud to communicate with the serial monitor in the Arduino IDE.

The second parameter is optional; it defines the data bits, parity, and stop bits of the serial protocol. If you omit the second parameter, the serial interface uses an 8-bit serial protocol, with no parity and 1 stop bit. This is the standard used by most serial communications devices. If the serial device you use requires a different setting, you can use labels to define the settings for the second parameter. For example, to use a 7-bit protocol with even parity and 1 stop bit, you use the label `SERIAL_7E1` for the second parameter.

Sending Data

Three separate functions send data to a remote device:

- `Serial.print`
- `Serial.println`
- `Serial.write`

The `Serial.print` function sends data as ASCII text. The ASCII format is commonly used for displaying data, which is what the Arduino IDE serial monitor uses. For numeric values, you can specify an optional second parameter, which defines the numeric format to use: `BIN` for output as a binary value, `DEC` for decimal format, `HEX` for hexadecimal format, and `OCT` for octal format. The default is decimal format.

The `Serial.println` function works the same as the `Serial.print` function, but adds the carriage return and line feed characters to the end of the output, creating a new line in the output window.

The `Serial.write` function allows you to send 1 byte of raw data to the remote device, without any formatting.

Receiving Data

The serial interface on the Arduino contains a buffer, holding data as the Arduino receives it on the RX pin. It can store up to 64 bytes of data, which allows you some flexibility in how your sketch retrieves the incoming data.

Your sketch can retrieve data from the buffer 1 byte at a time using the `Serial.read` function. Each time you read a byte from the buffer, the Arduino removes it from the buffer and shifts the remaining data over.

You can also retrieve multiple bytes of data from the buffer at a time. The `Serial.readBytes` function lets you specify the number of bytes to extract from the buffer and place them in an array variable in your sketch. The `Serial.readBytesUntil` function extracts data until a character that you specify is detected in the buffer data. That comes in handy if you send text data using a carriage return and line feed format for each line of data.

The `Serial.parseInt` or `Serial.parseFloat` functions provide ways for you to easily retrieve integer or floating-point values that you pass into your Arduino sketch as it runs. The next section shows an example of how to pass numeric values to your sketch and then retrieve them using the `Serial.parseInt` function.

Testing the Serial Port

Let's go through an example of sending data from the serial monitor to a running sketch to demonstrate how to use the receive features of the serial interface.

▼ TRY IT YOURSELF

Sending Data to Your Arduino

In this experiment, you control the blink rate of an LED by sending numeric values to your sketch using the Arduino IDE serial monitor. It uses the built-in LED connected to digital interface pin 13 on the Arduino, so you won't need to build an external circuit.

Here are the steps to create the code for the experiment:

1. Open the Arduino IDE.
2. Select Sketch from the menu bar, then select Import Library, and then select the Timer One library. (If you don't see the Timer One library listed, go to Hour 16, "Adding Interrupts," to see how to add it to the Arduino IDE library.)
3. Enter this code into the editor window:

```
#include <TimerOne.h>

int state = 0;
int value;
```

```
long int newtime;
void setup() {
  Serial.begin(9600);
  pinMode(13, OUTPUT);
  digitalWrite(13, state);
  Serial.println("Enter the blink rate:");
}

void loop() {
  if (Serial.available()) {
     value = Serial.parseInt();
     Serial.print("the blink rate is: ");
     Serial.println(value);
     Serial.println("Enter a new blink rate:");
     newtime = value * 1000000;
     Timer1.initialize(newtime);
     Timer1.attachInterrupt(blinkme);
  }
}

void blinkme() {
  state = !state;
  digitalWrite(13, state);
}
```

4. Save the sketch code as **sketch1701**.
5. Click the Upload icon to verify, compile, and upload the sketch to your Arduino unit.
6. Open the serial monitor utility to run the sketch and view the output. Make sure that the Line Speed drop-down box at the bottom is set for 9600 baud, and that the Line Ending drop-down box is set to No Line Ending.
7. In the text box at the top of the serial monitor, enter a number from 1 to 9, and then click the Send button.
8. Watch the LED blink at the rate that you specified in the serial monitor text box. You can submit new values to watch the rate change.

The sketch uses the `Serial.available` function to detect when new data is present in the serial interface buffer. It then uses the `Serial.parseInt` function to retrieve the data and convert it into an integer value. The sketch then uses the value in a timer interrupt (see Hour 16) to change how frequently the Arduino turns the LED on and off.

TIP

Serial Events

Reading data from the serial interface can sometimes be tricky, especially if you don't know exactly when to expect the data to arrive. Instead of using the `Serial.available` function to poll the serial interface for data, you can use serial events to notify your sketch when data is present.

Incoming serial data triggers a special function named `serialEvent`, which you can define in your sketch. You can include code in the `serialEvent` function to process the data as it comes in, and then return back to your normal sketch code, much like the interrupts shown in Hour 16.

Working with the SPI Port

The Serial Peripheral Interface (SPI) protocol uses a synchronous serial connection to communicate with one or more peripheral devices over a short distance. A synchronous serial connection requires a separate clock signal to synchronize the data transfer between devices. Many sensors use the SPI protocol to communicate with a host system. You can use the SPI port on your Arduino to interface with those types of sensors.

The SPI protocol uses a bus technology to share a single interface between multiple devices. One device connected to the bus is designated as the master and controls the operation of the bus. The other devices connected to the bus are designated as slaves and can send or receive signals on the bus only when polled by the master device.

WATCH OUT!

The Arduino and SPI Slave Mode

The Arduino hardware supports both master and slave mode in SPI, but at the time of this writing, the Arduino library only supports operating in master mode on the SPI bus. This means that you cannot currently use your Arduino to communicate with other SPI master devices, only SPI slave devices, such as sensors.

The following sections describe how to use your Arduino as an SPI master device to communicate with SPI slave devices, such as sensors.

The SPI Interfaces

The SPI protocol uses three signals that connect to all devices in the SPI bus:

- **MISO (Master In Slave Out):** Line for slave sending data to the master device.
- **MOSI (Master Out Slave In):** Line for the master device sending data to the slave.
- **SCK:** A serial clock signal to synchronize the data transfer.

Besides the three bus lines, each slave device has a Slave Select (SS) pin that connects to the master. The master must set the appropriate slave SS pin to a LOW value when it communicates with that specific slave device. If the SS pin is set to HIGH, the slave device ignores any data on the MOSI line.

The Arduino supports SPI signals using an interface that's separate from the standard header pins. All Arduino devices include a separate ICSP header on the far-right side of the unit, as shown in Figure 17.2.

FIGURE 17.2
The ICSP header on the Arduino Uno unit.

Figure 17.3 shows the pin location for the SPI signals on the ICSP header.

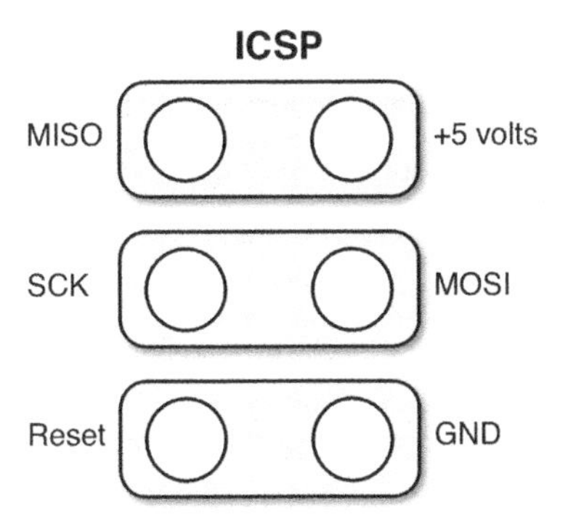

FIGURE 17.3
The SPI signals on the ICSP header.

The Arduino Uno also provides the SPI signals on the digital interface header pins:

- SS on digital interface 10
- MOSI on digital interface 11
- MISO on digital interface 12
- SCK on digital interface 13

If you use those interfaces to support SPI communication, you can't use them as digital inputs or outputs.

BY THE WAY

SPI and Digital Interfaces

When you initialize the SPI feature on the Arduino, you won't be able to use the digital interfaces assigned to the SPI signals as inputs or outputs. That includes the built-in LED connected to digital interface 13.

The SPI Library Functions

The Arduino IDE software includes a separate SPI library by default. The SPI library contains functions required to communicate with SPI slave devices using your Arduino, but doesn't provide any functions for the Arduino to act as a slave device itself. Table 17.2 lists the functions available in the SPI library.

TABLE 17.2 SPI Functions

Function	Description
`begin()`	Initializes the SPI interface pins on the Arduino.
`end()`	Terminates using the interface pins for SPI communication and returns them back to normal digital interface mode.
`setBitOrder(mode)`	Sets how the Arduino sends data out the SPI interface.
`setClockDivider(divider)`	Sets the timing clock speed on the SCK interface.
`setDataMode(mode)`	Sets the SPI data mode on the interface.
`transfer(val)`	Sends 1 byte of data out the SPI bus, and retrieves 1 byte of data from the bus.

The SPI bus transfers data 1 byte at a time. You must use the `SPI.setBitOrder` function to set whether the Arduino handles the byte in least-significant bit order (`LSBFIRST`) or most-significant bit order (`MSBFIRST`).

The `SPI.setClockDivider` function determines the clock speed set for the SPI bus. The clock speed on the master and slave devices must match or you won't get the proper data synchronization. The SPI library allows you to set the SPI bus clock speed to even number divisions of the Arduino microcontroller clock, which is 16MHz. You specify the clock speed using labels, such as `SPI_CLOCK_DIV2` to set the clock speed to 8MHz (half of the Arduino 16MHz clock speed), or `SPI_CLOCK_DIV4` to set the clock speed to 4MHZ. The parameter supports values of 2, 4, 8, 16, 32, 64, or 128.

The `SPI.setDataMode` function sets the clock polarity and phase used in the SPI bus. The SPI bus can use four standard data modes. All devices on the bus must be set to use the same mode. The values you can use are `SPI_MODE0`, `SPI_MODE1`, `SPI_MODE2`, and `SPI_MODE3`. You will need to consult with the specifications for the SPI device you're communicating with to determine the mode it uses and set the Arduino to use the same SPI mode.

After you have all the SPI bus parameters set, you're ready to send and receive data on the SPI bus. The `SPI.transfer` function both sends and receives a single byte of data on the bus with one function. The `SPI.transfer` function pulls the SS interface low, so the slave device connected to that pin knows to read the data that the Arduino sends on the bus. This means you can only communicate with one SPI device from most Arduino models. The Arduino Due supports three SPI interfaces, allowing you to connect up to three SPI slave devices.

Working with I^2C

The I^2C protocol was developed by the Phillips Semiconductor Corporation for providing a communication protocol between multiple embedded electronic devices using only a two-wire bus system. The I^2C protocol is intended for very short distances, often between devices placed on the same circuit board, but can also be used with short distance wired connections.

To eliminate the extra slave select line, the I^2C protocol uses addresses to determine which data is intended for which device. Each slave device on the bus has a unique address assigned to it, and only responds to data sent to its own address, much like an Ethernet local-area network (LAN).

This section discusses the I^2C support provided by the Arduino and demonstrates how to communicate between Arduino devices using the I^2C protocol.

The I^2C Interface

One of the selling features of the I^2C protocol is that it requires only two signals:

- **Serial data line (SDA):** Sends the data between devices.
- **Serial clock (SCL):** Provides a clock signal to synchronize the data transfer.

The single data line is used to both send and receive data, so it's important that the master device have full control over the bus at all times to prevent data collisions. Slave devices can only send data when prompted by the master device.

Unfortunately, the different Arduino models provide the two I^2C signals on different interface pins, so you have to be careful when accessing the I^2C signals on your specific Arduino unit. Table 17.3 shows where to find the I^2C signals on the different Arduino models.

TABLE 17.3 I^2C Interface Pins

Model	SDA	SCL
Uno	A4	A5
Due	20	21
Ethernet	A4	A5
Leonardo	2	3
Mega	20	21

Notice that the Uno and Ethernet models use two analog interface pins for the I^2C signals instead of digital interface pins. The Due Arduino model also supports a second dedicated I^2C interface pair. Those pins are labeled SDA1 and SCL1 on the board.

The Wire Library Functions

The Wire library provides software support for using the I^2C interface on the Arduino. The Arduino IDE includes the Wire library by default, so it's easy to use in your sketches. The Wire library provides all the functions that you need to set up and work with the I^2C protocol in your sketches. Table 17.4 shows the functions that are available.

TABLE 17.4 Wire Functions

Function	Description
`available()`	Returns the number of bytes available for retrieval from the buffer.
`begin([address])`	Initialize the I^2C interface. If an address is specified, the Arduino joins as a slave device. If an address is omitted, the Arduino joins as a master device.
`beginTransmission(address)`	Start a transmission session to the slave device specified by *address*.
`endTransmission(address)`	Ends the transmission session with the slave device specified by *address*.
`read()`	Read a byte of data sent from a master device.
`requestFrom(address, bytes)`	Request *bytes* number of data bytes from the slave device specified by *address*.
`onReceive(function)`	Calls *function* when the Arduino receives data from a master device.
`onRequest(function)`	Calls *function* when the Arduino receives data from a slave device.
`write(data[, length])`	Send data in the transmission session with a slave device. The *data* parameter can be a single byte of data, or a null-terminated string value. If you specify *data* as an array, you must also include the *length* of the array as the second parameter.

Every slave device on the I^2C bus requires a unique address. To communicate on the bus as a slave device, you must assign your Arduino an address, specified in the `begin` function:

```
Wire.begin(1);
```

This statement sets the address of the Arduino to 1. Be careful when you assign an address to your Arduino that it doesn't conflict with the address assigned to any other slave devices on the I^2C bus. To operate as a master device, just use the `begin` function without specifying an address.

After the `begin` function, your Arduino can send and receive data messages from other devices on the I^2C bus. Sending data to a device requires three separate statements. First, you must use the `beginTransmission` function to identify the slave device the data is intended for. Next, you use one or more `write` functions to send the actual data. Finally, you use the `endTransmission` function to tell the slave device you're done talking to it.

Receiving data depends on whether the Arduino is operating in master or slave mode. If in slave mode, use the `read` function to retrieve the data sent from the master device on the bus. If the Arduino is operating in master mode, you need to use the `requestFrom` function to specify the address of the slave device to retrieve data from.

On the surface, using the I^2C protocol can look somewhat complicated, but once you get the hang of the master and slave modes, it's a breeze to send and receive data. The following section goes through an example of using the I^2C protocol to communicate between two Arduino units.

Testing the I^2C Interface

One great feature of the I^2C bus is that you can use it to communicate between multiple Arduino units. You can then use one Arduino as the master and connect the others as slaves on the I^2C bus. The master can request sensor data from each of the slaves to combine readings for logging or display purposes.

If you have two Arduino units handy (they don't have to be the same model), you can work through this experiment.

▼ TRY IT YOURSELF

Communicating Between Arduino Units

In this experiment, you set up an I^2C bus to connect two Arduino units together so one Arduino can control the actions of another Arduino. The master Arduino unit will listen on the serial interface for an integer value and then pass that value to the slave Arduino using the I^2C bus, which will use it to control the blink rate of the LED on digital interface 13.

Connecting two Arduino units sounds easy, but unfortunately you cannot just connect the I^2C pins on the Arduino units directly together. The I^2C bus protocol requires that the SDA and SCL lines be held at a HIGH voltage level when there isn't any data on them. To do that, you need some equipment to set the pullup resistors:

- Two 1K-ohm resistors (color code brown, black, red)
- Eight jumper wires
- A standard breadboard

First, follow these steps to create the circuit to connect the two Arduino units:

1. Place the two 1K-ohm resistors on the breadboard so that one lead of each resistor connects to a common bus on the breadboard and the other leads connect to separate rails. The rails will carry the SDA and SCL bus signals.
2. Connect the common bus that has the two resistor leads on the breadboard to the 5V pin on the master Arduino using a jumper wire.

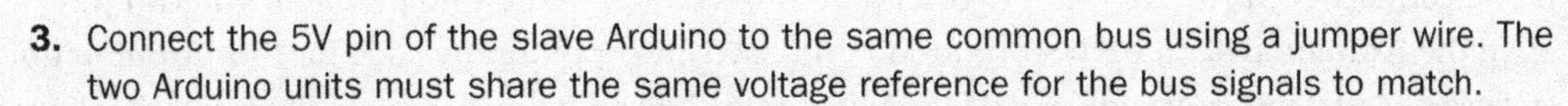

3. Connect the 5V pin of the slave Arduino to the same common bus using a jumper wire. The two Arduino units must share the same voltage reference for the bus signals to match.
4. Connect the GND pin of the master Arduino to the GND pin of the slave Arduino using a jumper wire. The two Arduino units must also share the same ground reference for the bus signals to match.
5. Connect the SDA signal pin on each Arduino (for the Uno, analog pin A4) to one of the bus rails on the breadboard that has the 1K-ohm resistor connected to it.
6. Connect the SCL pin on each Arduino (for the Uno, analog pin A5) to the other rail on the breadboard that has the 1K-ohm resistor on it.

Figure 17.4 shows the circuit diagram for this connection.

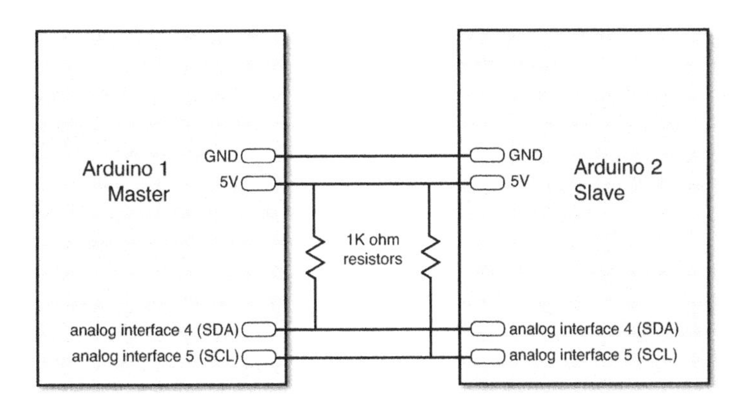

FIGURE 17.4
Circuit diagram for connecting the I^2C bus pins on the Arduino units.

WATCH OUT!

Mismatched Arduinos

Be careful if you're not using the same model of Arduino unit for both devices. The Arduino Due model only supports 3.3 volts, so you'll have to use the 3.3V pin on the Arduino Uno if you want to connect it to an Arduino Due.

Next comes the tricky part—setting up the code for the master and slave Arduino units. Because each Arduino unit runs a different sketch, you must be careful when you connect the Arduino IDE that you're loading the correct sketch into the correct Arduino.

If you connect each Arduino unit to a separate workstation, there's no problem; you can just open the Arduino IDE on each workstation and enter the appropriate code for that Arduino unit. However, if you just have one workstation, you must be careful as to which one is which when connecting the Arduino units.

The easiest way to do that is to just connect one Arduino unit at a time to the workstation, program it, and then connect the other Arduino to program it.

First, connect the slave Arduino unit to the USB port, and then follow these steps:

1. Open the Arduino IDE.
2. Select Sketch from the menu bar, then Import library, and then select the Wire library.
3. Select Sketch again from the menu bar, then Import Library, and then select the Timer One library.
4. Enter this code into the editor window (the two `#include` directives should already be present from the import):

```
#include <Wire.h>
#include <TimerOne.h>

int state = 0;
int value;
long int newtime;

void setup() {
  Wire.begin(1);
  Wire.onReceive(gotRate);
  pinMode(13, OUTPUT);
  digitalWrite(13, state);
}

void loop() {
}

void gotRate(int howMany) {
  if (Wire.available())
  {
    value = Wire.read();
    newtime = value * 1000000;
    Timer1.initialize(newtime);
    Timer1.attachInterrupt(blinkme);
  }
}
```

```
void blinkme() {
  state = !state;
  digitalWrite(13, state);
}
```

5. Save the slave sketch as **sketch1702**.
6. Click the Upload icon to verify, compile, and upload the sketch to the slave Arduino.

Now you're ready to code the master Arduino unit. Disconnect the slave Arduino from the USB interface, and plug the master Arduino unit in. Follow these steps to code it:

1. Open the Arduino IDE.
2. Select Sketch from the menu bar, then Import Library, and then select the Wire library.
3. Enter this code into the editor window:

```
#include <Wire.h>

int value;
void setup() {
  Serial.begin(9600);
  Wire.begin();
  Serial.println("Enter the blink rate:");
}

void loop() {
  if (Serial.available()) {
     value = Serial.parseInt();
     Serial.print("the blink rate is: ");
     Serial.println(value);
     Wire.beginTransmission(1);
     Wire.write(value);
     Wire.endTransmission(1);
     Serial.println("Enter a new blink rate:");
  }
}
```

4. Save the code as **sketch1703**.
5. Click the Upload icon to verify, compile, and upload the code to the master Arduino unit.
6. Open the serial monitor to run the master program. Because you connected the 5V and GND pins of the two Arduino units together, the slave Arduino unit will receive power from the master Arduino unit plugged into the USB port. It might help to press the Reset button on the slave Arduino to ensure that it has restarted the slave sketch code properly.

7. Enter a numeric value from 1 to 9 in the text box at the top of the serial monitor, and click the Send button to send it to the master Arduino unit. The LED on the slave Arduino unit should begin to blink at the rate you entered.

The master Arduino unit uses the I^2C bus to communicate the integer value you entered to the slave Arduino unit, which then uses it to set the `TimerOne` function interval to make the LED on digital interface 13 blink. That's a pretty cool experiment!

Summary

This hour discussed how to communicate from your Arduino with other devices using serial protocols. The Arduino supports three separate serial protocols: the standard serial interface, the Serial Peripheral Interface (SPI), and the Inter-integrated Circuit (I^2C) protocol. The standard serial interface uses one wire to send and another to receive data bits. The Arduino also includes a built-in serial-to-USB convertor that enables you to access the serial port using a USB cable. You can access the serial port using the Serial library provided in the Arduino IDE. The SPI protocol is available from the ICSP interface on the Arduino and uses three wires to communicate with external devices. You use the SPI library to access functions to send and receive data using the SPI port. Finally, the Arduino supports the I^2C protocol using two interface ports and functions from the Wire library.

In the next hour, you'll see how to interface your Arduino with different types of sensors for monitoring various conditions, including light, sound, and motion.

Workshop

Quiz

1. What Arduino IDE library should you use to work with the I^2C interface on the Arduino?

   A. The SPI library

   B. The Wire library

   C. The Serial library

   D. The Interrupt library

2. You can connect the Arduino serial interface directly to a standard COM interface used in PC workstations. True or false?

3. What makes the I^2C protocol different from the SPI protocol?

Answers

1. B. The Wire library contains the functions required to communicate with the I^2C interface on the Arduino.
2. False. The COM interface used in Windows workstations uses a 12V signal reference, which is too large for the TTL-level serial port interface on the Arduino.
3. The I^2C protocol assigns unique addresses to each slave device on the bus. The master device can communicate with a specific device by specifying its address.

Q&A

Q. How many sensors can an Arduino control on a single I^2C bus?

A. The I^2C protocol uses 7-bit addresses for slave devices, allowing you to assign addresses from 0 to 127. That means you can have up to 128 sensors on a single I^2C bus.

Q. When the Arduino operates as a master device in an SPI bus, how does it communicate with more than one slave device? There's only one SS pin that the master can control.

A. The Arduino can use any other available digital interface port for the SS pin to control a slave device. Just ensure that the port is held at a HIGH level by default and that the Arduino sets it to a LOW level when trying to send data to the slave device.

HOUR 18
Using Sensors

What You'll Learn in This Hour:

- The different types of analog sensors available
- How to use voltage-based sensors
- How to work with resistance-based sensors
- How touch sensors work

Working with the analog interfaces on the Arduino can be somewhat complex. There are lots of different analog sensors out there, as well as lots of different ways to measure the analog data they produce. This hour takes a closer look at how to work with the different types of analog sensors that you may run into when working with your Arduino.

Interfacing with Analog Sensors

Analog sensors are more difficult to work with than digital sensors. With digital sensors, you only have to worry about detecting one of two possible values (a 0 or a 1). However, analog sensors can produce an infinite number of possible output values. The trick to working with analog sensors is in knowing what type of output the sensor produces and decoding just what the output means.

To convey information using an analog signal, analog sensors typically change one of three different physical electrical properties:

- Voltage
- Resistance
- Capacitance

You must know the type of analog sensor that your circuit uses to be able to appropriately handle the analog output that it generates. Also, your sketch will need to know how to convert the analog value to meaningful information, such as the temperature or the brightness level.

With analog signals, your Arduino sketch will often need help from some external electronic circuitry to change the signal the sensor produces into some type of usable form. The combination of the electronic circuit and your sketch is what creates the monitoring environment.

The next sections discuss how to monitor each of the three different types of analog sensors you may have to work with.

Working with Voltage

Many analog sensors produce a varying voltage that represents the property they monitor. There are different types of analog sensors that use voltage as the output used in industry, such as the following:

- Temperature sensors
- Light sensors
- Motion sensors

With each of these sensors, the amount of temperature, light, or motion that the sensor detects is represented by the voltage level present at the sensor output.

The good thing about using voltage-based analog sensors is that because the Arduino analog interfaces detect changes in voltage, you often don't need to provide too much additional electronic circuitry to work with voltage-based sensors. This section covers the different issues you need to consider when working with analog voltages from sensors.

Voltage Range

The first thing to consider when working with voltage-based analog sensors is the voltage range the sensor produces. An upper and lower limit usually applies to the voltage that the sensor produces under the monitoring conditions. For example, the Arduino Uno uses a 5V reference voltage, so the analog input interfaces are based on the maximum sensor voltage being 5V.

However, not all analog sensors are based on a 5V reference voltage. You might have to manipulate the output voltage from the sensor to be able to work with it in the Arduino.

WATCH OUT!

The Arduino Due

Be careful when using the Arduino Due, Fio, and Pro; they support only 3.3V voltage levels.

Using a Sensor That Matches the Arduino Voltage

The simplest situation is when the sensor outputs a signal that matches the voltage the Arduino can support (3.3V for the Due, or 5V for the other models). In this case, you can just connect the sensor output directly to the Arduino analog input interface, as shown in Figure 18.1.

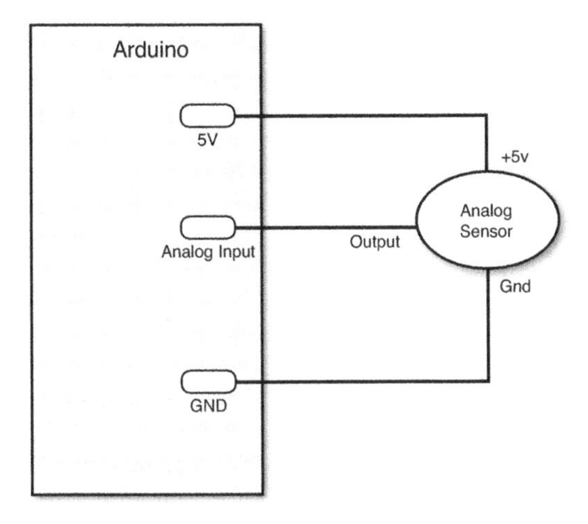

FIGURE 18.1
Connecting a 5V sensor to the Arduino Uno.

Notice from Figure 18.1 that you can also power the sensor directly from the Arduino Uno 5V pin, which provides 5V of power. You will also need to connect the ground for the sensor to the Arduino GND pin to ensure the circuit is complete.

Using Larger Voltages

If the sensor output voltage is larger than what the Arduino supports, you cannot directly connect the sensor output to the analog input interface; otherwise, you may damage your Arduino. Instead, you need to use what's called a voltage divider circuit to reduce the amount of voltage present on the analog interface.

A voltage divider is two resistors placed in series, with the output taken from the connection between the two resistors, as shown in Figure 18.2.

The output of the voltage divider depends on the ratio of the two resistor values. The equation that determines the output voltage is as follows:

```
output voltage = sensor voltage * (R2 / (R1 + R2))
```

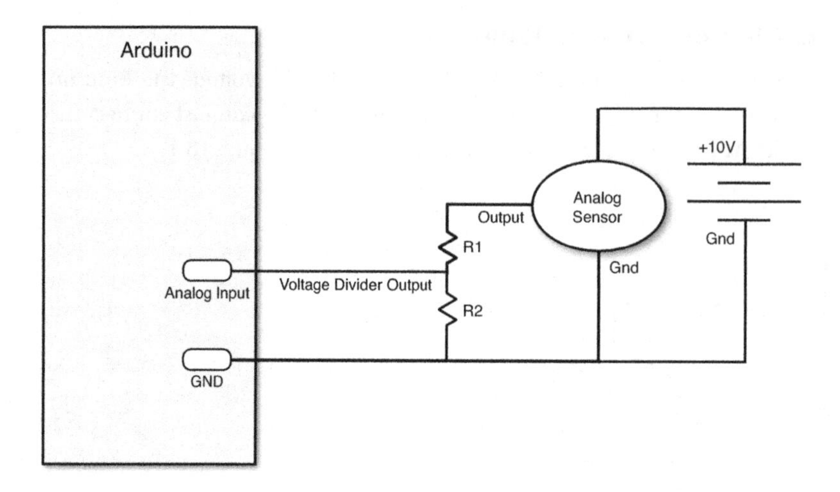

FIGURE 18.2
Using a voltage divider on the analog input interface.

If both resistors in the voltage divider are the same value, the output voltage will be one half of the sensor voltage. So, if the analog sensor has a maximum output of 10V, you can use two 1K-ohm resistors to ensure that the maximum voltage present at the analog input is less than 5V.

WATCH OUT!

Common Ground

Whenever you work with a sensor that uses its own power supply, make sure that you connect the ground from the external power supply to the GND pin on the Arduino unit. This ensures the sensor and the Arduino are using the same ground reference. However, don't connect the power pole of the external power supply to the Arduino 5V pin!

Using Smaller Voltages

The opposite of too much voltage is too little voltage to detect. If the sensor outputs a small voltage, you can connect the sensor directly to the Arduino analog input interface, but you may have trouble reading the changes in the output voltage. The solution to that is to change the voltage reference that the Arduino uses to represent the maximum output value.

To change the voltage reference, you must use the AREF input pin on the Arduino. Just connect the voltage source from the sensor to the AREF pin on the Arduino, as shown in Figure 18.3.

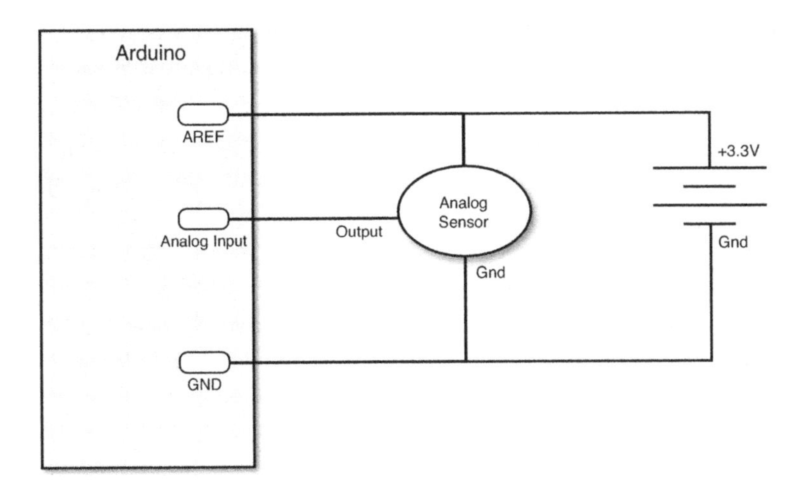

FIGURE 18.3
Changing the reference voltage used for the Arduino analog input.

With the new reference voltage, the Arduino changes the output range to represent the new reference voltage instead of 5V. To get your sketch to use the new reference voltage, you must use the `AnalogReference` function:

```
AnalogReference(EXTERNAL);
```

When you use the `EXTERNAL` setting, the `analogRead` function bases its output for a maximum voltage equal to the reference voltage. After you set the analog reference voltage, the `analogRead` function returns the 1023 value when the voltage on the input interface is the same as the reference voltage.

Sensitivity

Another important issue to consider is the sensor sensitivity. Your Arduino sketch must be able to detect the changes in the voltage sent from the analog sensor, which represent changes in the property that's being monitored.

The Arduino Uno uses a 10-bit analog-to-digital converter (ADC), which provides 1024 separate output values based on the input voltage level. If your sensor uses 5V as reference, that means the Arduino will produce a value of 0 when the input voltage is 0V, and 1023 when the input voltage is 5V.

By dividing that voltage range, you'll notice that each step in the output represents about a 0.005V change in the input voltage, or about 5 millivolts (mV). That means the Arduino analog input won't be able to distinguish voltage changes less than 5mV from the sensor.

However, you may also run into the situation where the Arduino analog input may be too sensitive to changes in the voltage received from the sensor. You may not want your sketch to respond to a 5mV change in the voltage, but rather, only to larger changes.

One solution to that problem is to remap the output range generated by the analogRead function to a smaller range using the map function (see Hour 15, "Interfacing with Analog Devices"). The map function allows you to specify a new range to use for the output:

```
input = map(analogRead(A0), 0, 1023, 0, 255);
```

This desensitizes the output from the analogRead function, providing for fewer changes within the range of the sensor. By changing the range to 0 through 255, the input value only changes for changes of 20mV instead of 5mV.

Converting Voltage to Something Useful

The last step in using a voltage-based sensor is to convert the voltage that the Arduino detects to a meaningful value that represents the property that you're monitoring (such as the temperature of the room, or the amount of light that's present). Usually this requires having to do a little research on the sensor that you're using.

Most sensor manufacturers produce a datasheet that documents the range of output of the sensor and the relation to the property that it measures. For example, the TMP36 temperature sensor produces a voltage range from 0.1V for –40 degrees Celsius to 2.0V for 150 degrees Celsius. To display the actual temperature, you'll need to do a little math in your Arduino sketch. According to the TMP36 datasheet, the relation of the output voltage to the actual temperature uses this equation:

```
temp in Celsius = (voltage - 500) / 10
```

Where the *voltage* value is specified in millivolts.

However, before you use that equation, you must convert the integer value that the analogRead function returns into a millivolt value. A little bit of algebra can help out here.

You know that for a 5000mV (5V) value span the analogRead function will return 1024 possible values. Just use an equation to relate the actual analogRead output value to the known voltage values, and then solve for the voltage variable:

```
voltage / output = 5000 / 1024
voltage = (5000 / 1024) * output
```

Using this equation, you can take the *output* value you retrieve from the analogRead function and determine the voltage present at the analog input. Once you know the voltage, you can use it to find the temperature. Or, if you prefer, you can combine both equations into a single calculation:

```
tempC = ((5000/1024) * output) - 500) / 10;
```

Now you can retrieve the temperature directly from the *output* integer value the `analogRead` function produces. The next section walks through an example of using this process to determine the temperature using the TMP36 temperature sensor.

Using a Voltage-Based Sensor

This experiment uses the popular TMP36 temperature sensor to detect the temperature and display that temperature in the serial monitor. The TMP36 sensor is common in the Arduino world. You'll often find it in Arduino kits, including the official Arduino Starter Kit package. It's also readily available for purchase from many Arduino electronic suppliers such as Adafruit and Newark Electronics.

TRY IT YOURSELF ▼

Detecting Temperature

In this experiment, you need to connect the TMP36 directly to your Arduino unit. The TMP36 temperature sensor has three leads: power, ground, and the output.

You must power the sensor by connecting the power lead to a voltage source between 2.7 and 5.5V (which makes it ideal for use with both the 5V Arduino Uno and the 3.3V Arduino Due) and connect the ground lead to the Arduino ground. The middle lead produces an output voltage less than 5V, so you can connect that directly to an analog input pin. Figure 18.4 shows the pin outputs on the TMP36 temperature sensor.

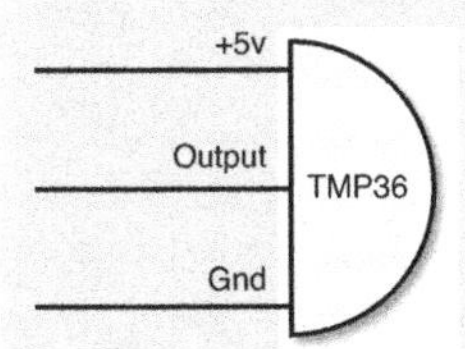

FIGURE 18.4
The TMP36 sensor pins.

Here are the steps to wiring the TMP36 sensor to your Arduino unit:

1. Place the TMP36 sensor on the breadboard so that each lead connects to a separate rail section and that the flat side of the TMP36 case is facing toward the left.

2. Place a jumper wire between the top lead of the TMP36 and the 5V pin on the Arduino.
3. Place a jumper wire between the bottom lead of the TMP36 and the GND pin on the Arduino.
4. Place a jumper wire between the middle lead of the TMP36 and the Analog 0 pin on the Arduino.

That's all the hardware you need to worry about for this experiment. The next step is to create the sketch code. Just follow these steps to work on that:

1. Open the Arduino IDE and enter this code into the editor window:

```
int output;
float voltage;
float tempC;
float tempF;

void setup() {
   Serial.begin(9600);
}

void loop() {
   output = analogRead(A0);
   voltage = output * (5000.0 / 1024.0);
   tempC = (voltage - 500) / 10;
   tempF = (tempC * 9.0 / 5.0) + 32.0;
   Serial.print("Sensor Output: ");
   Serial.println(output);
   Serial.print("Voltage (mv): ");
   Serial.println(voltage);
   Serial.print("Temp (C): ");
   Serial.println(tempC);
   Serial.print("Temp (F): ");
   Serial.println(tempF);
   Serial.println();
   delay(5000);
}
```

2. Save the sketch as **sketch1801**.
3. Click the Upload icon to verify, compile, and upload the sketch code into your Arduino unit.
4. Open the serial monitor to view the output from the sketch.

The output from the temperature sensor should be somewhat consistent when at room temperature. Try placing your fingers around the sensor and see whether the temperature rises. Then, place an ice cube in a plastic bag, and then place the bag next to the sensor. That should lower the output generated by the sensor.

Working with Resistance Output

Instead of using a voltage output, some analog sensors change their resistance as the characteristic they monitor changes. These types of sensors include the following:

- **Thermistors:** Change resistance due to temperature.
- **Photoresistors:** Change resistance due to light.

The problem with resistance sensors is that the Arduino analog interfaces can't directly detect resistance changes. This will require some extra electronic components on your part.

The easiest way to detect a change in resistance is to convert that change to a voltage change. You do that using our friend the voltage divider, as shown in Figure 18.5.

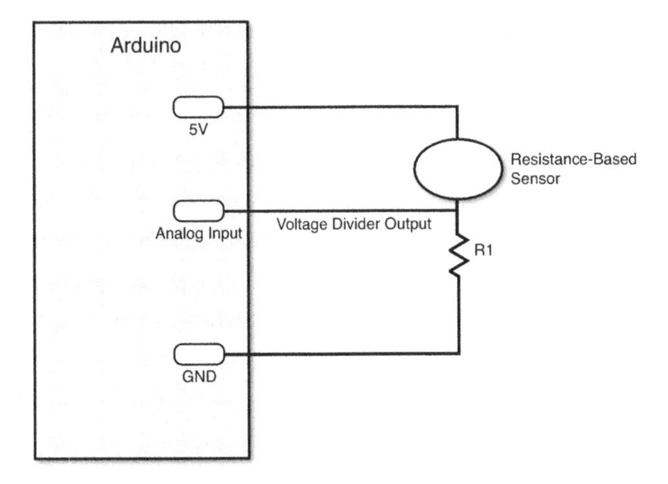

FIGURE 18.5
Using a voltage divider to detect a change in resistance.

By keeping the power source output constant, as the resistance of the sensor changes, the voltage divider circuit changes, and the output voltage changes. The size of resistor you need for the R1 resistor depends on the resistance range generated by the sensor and how sensitive you want the output voltage to change. Generally, a value between 1K and 10K ohms works just fine to create a meaningful output voltage that you can detect in your Arduino analog input interface.

The voltage divider will produce a varying output voltage. As the sensor resistance value increases, the output voltage increases, and as the sensor resistance decreases, the output voltage decreases. At that point, you can use the same sensing tricks that you learned for working with the voltage sensors.

Using a Resistance-Based Sensor

So, the key to using resistance-based sensors is to build a voltage divider circuit to generate the output voltage. The following experiment demonstrates how to do that using a photoresistor to detect light levels.

▼ TRY IT YOURSELF

Try It Yourself: Building a Light Meter

In this experiment, you use a common photoresistor (also called a photocell or a light-dependent resistor (LDR)) to detect the amount of light in the room. Photoresistors come in many shapes and sizes. Any type will work for this experiment.

The photoresistor uses two output leads and changes the resistance present on those leads as the light level changes. As the amount of light increases, the resistance of the photoresistor decreases (and vice versa). Many standard Arduino kits come with one or more photoresistors, including the Arduino Start Kit, and they're also available for purchase from the standard Arduino electronics shops.

Here are the steps for creating the Arduino light meter project. First, you need to build the electronic circuit:

1. Place the photoresistor leads on two separate rails sections on the breadboard.
2. Connect a 10K-ohm resistor (color code brown, black, orange) so that one lead connects to one lead of the photoresistor and the other lead is in another rail on the breadboard.
3. Connect the 5V pin of the Arduino to the free lead on the photoresistor.
4. Connect the GND pin of the Arduino to the free lead on the 10K-ohm resistor.
5. Connect the Analog 0 interface pin on the Arduino to the rail that has both the photoresistor and 10K-ohm resistor leads.

The circuit creates a standard voltage divider, sending the output voltage to the Analog 0 interface pin on the Arduino.

The next step is to create a sketch to read the output values. Follow these steps to do that:

1. Open the Arduino IDE and enter this code into the editor window:

```
int output;
float voltage;

void setup() {
   Serial.begin(9600);
}

void loop() {
   output = analogRead(A0);
   voltage = output * (5000.0 / 1024.0);
   Serial.print("Sensor Output: ");
   Serial.println(output);
   Serial.print("Voltage (mv): ");
   Serial.println(voltage);
   Serial.println();
   delay(5000);
}
```

2. Save the sketch as **sketch1802**.
3. Click the Upload icon to verify, compile, and upload the sketch to your Arduino unit.
4. Open the serial monitor to run the sketch and view the output.

Check the output from the sketch using the normal room lighting, and then try covering the photoresistor with your hand and watch the output. The sensor output value should decrease, indicating a decrease in the voltage present at the analog input.

As the photoresistor receives less light, the resistance value increases, causing less voltage to be output from the voltage divider. As you increase the light on the photoresistor, the resistance value decreases, causing more voltage to be output from the voltage divider.

Using Touch Sensors

Touch sensors use capacitance to detect when an object is being touched. When you touch a metal object, the natural capacitance in your body changes the capacitance present in the object.

The problem with using touch sensors is similar to the problem we ran into with resistance-based sensors; they both don't directly change output voltage. However, detecting the change in capacitance from a touch sensor is solved using an interesting electrical property.

Capacitors store voltage. The amount of time it takes for a capacitor to fully charge or discharge depends on the amount of resistance present in the circuit, as well as the capacitance of the capacitor. By setting the resistor and capacitor sizes, you can control just how quickly the capacitor charges and discharges. This is called an RC circuit.

You can build a simple RC circuit using the capacitance detected by a touch sensor and a resistor of a known value. Because the resistance value is known, you can determine the capacitance value based on the time it takes for the capacitor to charge. You can tell when a capacitor is fully charged when current starts following past the capacitor in the circuit (until the capacitor is fully charged, no current will flow past it).

So basically, detecting the capacitance change has to do with timing. Just apply a known voltage value to one end of the RC circuit and detect how long it takes for the same voltage to be present at the other end of the circuit. The more capacitance, the longer it takes to detect the output voltage.

To do that, you use two digital input pins on the Arduino. (That's right, you use digital pins and not analog pins for this.) You set one pin as output to a HIGH value, and then track the amount of time it takes for the other digital input pin to have a HIGH value. The longer it takes, the more capacitance is in the circuit, and most likely, someone is touching the sensor.

Working with Touch Sensors

Fortunately for us, some smart developers have already created an Arduino library to easily detect the timing change between the two digital output pins for us. The CapacitiveSensor library includes all the functions we need to build a capacitive touch sensor project.

This experiment uses the CapacitiveSensor library to detect when a wire is touched.

▼ TRY IT YOURSELF

Building a Touch Sensor

You can build a simple touch sensor using just a resistor, some wire, and your Arduino unit. First, follow these steps to build the circuit:

1. On a breadboard, plug a 1M (mega)-ohm resistor (color code brown, black, green) between two separate rails.
2. Use a jumper wire to connect one lead of the resistor to digital pin 7. This is the Send pin for our circuit.
3. Use a jumper wire to connect the other lead of the resistor to digital pin 5. This is the Receive pin for our circuit.
4. Connect a wire to the rail that contains the digital pin 5 connection, and let the other end of the wire be free.

That completes the touch sensor circuit. The free end of the wire is our sensor. Some experiments also use aluminum foil connected to the end of the wire. Feel free to try that as well. Figure 18.6 shows the diagram for the completed circuit.

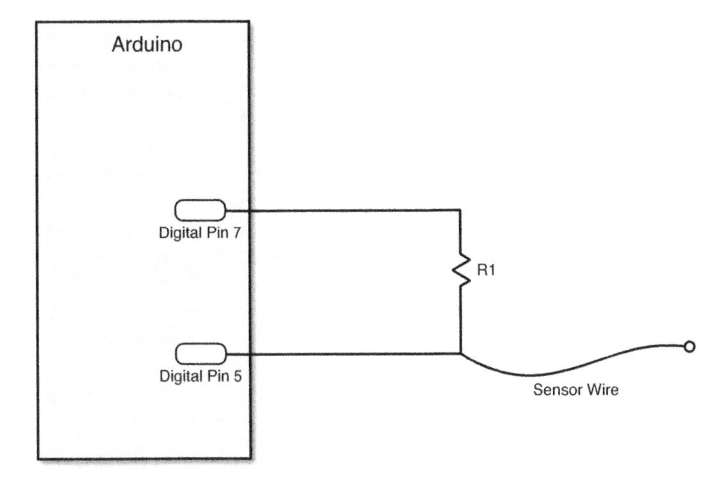

FIGURE 18.6
The touch sensor circuit.

Now you're ready to build the sketch code. Here are the steps to do that:

1. Download the CapacitiveSensor library from the link on the Arduino Playground website (http://playground.arduino.cc//Main/CapacitiveSensor).
2. Extract the CapacitiveSensor folder from the downloaded zip file and place it in your Arduino libraries folder, usually located in your Documents folder for both Windows and Apple OS X systems.
3. Open the Arduino IDE, and from the menu bar, select Sketch, Import Library, CapacitiveSensor.
4. In the IDE editor window, enter this code:

```
#include <CapacitiveSensor.h>

CapacitiveSensor Sensor = CapacitiveSensor(7,5);

void setup() {
  Serial.begin(9600);
}
```

```
void loop() {
  long sensorValue = Sensor.capacitiveSensor(30);
  Serial.println(sensorValue);
  delay(2000);
}
```

5. Save the sketch as **sketch1803**.
6. Click the Upload icon to verify, compile, and upload the sketch to your Arduino unit.
7. Open the serial monitor to run the sketch and view the output.

Watch the sensor values that appear in the output when the sensor wire is untouched. Touch the end of the sensor wire and note the change in the sensor value. Next, hold the end of the sensor wire in your fingers and note the change in the sensor value.

TIP

Sensor Sensitivity

You may have to change the sensitivity of the sensor by changing the resistance value of the R1 resistor in the circuit for it to detect the touch. Higher resistance values make the sensor more sensitive to capacitance changes on the wire.

Summary

This hour discussed how to work with different types of analog sensors in your Arduino projects. Voltage-based sensors change the output voltage based on the property they monitor. You may have to limit the voltage present on the Arduino analog input pin using a voltage divider, or change the sensitivity of the input pin using an external reference voltage and the `analogReference` function. Resistance-based sensors change their resistance as the monitored property changes. To use these types of sensors, you'll need to create a voltage divider to convert the resistance change to a voltage change. This chapter also covered how to use touch sensors in your Arduino circuits. Touch sensors are based on capacitance changes. To detect the capacitance change, you can build a simple RC circuit and use the CapacitiveSensor library to detect the time change required to trigger a digital input pin.

In the next hour, we'll turn our attention to another popular Arduino circuit topic: motors. You can use different types of motors in your Arduino projects, and you need to know how to work with each type.

Workshop

Quiz

1. What kind of circuit should you use to decrease the voltage from the sensor?

   A. A voltage divider

   B. An RC circuit

   C. Change the Arduino reference voltage

2. You can connect the output of a resistor-based sensor directly to the Arduino analog interface to detect the sensor output change. True or false?

3. What type of circuit do touch sensors require to detect touch?

Answers

1. A. The voltage divider circuit allows you to decrease a voltage to a value within the range required for the Arduino.

2. False. For a resistance-based sensor, you must use a voltage divider to change the output voltage relative to the change of the sensor resistance.

3. An RC circuit provides a way to detect a change in the capacitance of a touch sensor by measuring the amount of time it takes for the capacitor to charge.

Q&A

Q. Can you use a voltage divider to decrease voltages generated by high-power devices?

A. Yes, but you may lose some sensitivity. By significantly decreasing the voltage, you won't be able to detect small voltage changes from the original voltage.

Q. Can the Arduino analog interface use sensors that produce AC voltages?

A. Not directly. The Arduino analog interfaces use DC voltages, so you first have to convert the AC output voltage to a DC voltage.

HOUR 19

Working with Motors

What You'll Learn in This Hour:

- ▶ The different types of motors you can use
- ▶ How to control motor speed
- ▶ How to control motor direction
- ▶ How to work with servo motors

With the growing popularity of mechanical automation, at some point you may end up working with motors in your projects. The Arduino is excellent for controlling motors, from simple mechanical decorations to complex industrial robots. This hour takes a closer look at how to work with the different types of motors that you may run into when working with your Arduino projects.

Types of Motors

Plenty of different types of motors are available to work with in your Arduino experiments. The type of motor you select usually depends on just what you need to move and how you need to move it. The three most common that you'll run into, though, are as follows:

- ▶ DC motors
- ▶ Stepper motors
- ▶ Servo motors

The following sections discuss the difference between these motors in detail.

DC Motors

DC motors consist of a solid shaft with wire wrapped around it, surrounded by two magnets. DC motors are so named because they run on direct current (DC) voltage, which provides current to the wires wound around the shaft, which creates a magnetic field, making the shaft rotate within the fixed magnets around the shaft.

Brushes are used to connect the wires wound around the shaft to the external circuit. The brushes allow the shaft to rotate and still remain connected to the electrical circuit. Because of this, you'll often see the term *DC brushed motors* used.

With DC motors, the more voltage you apply to the motor, the faster it spins, up to the maximum voltage that the motor supports. The nice thing about DC motors is that to get the motor to turn in the opposite direction, you just apply voltage in the opposite polarity to the wires.

Stepper Motors

A stepper motor is a special application of a DC motor. Instead of rotating the shaft at a constant speed, the stepper motor uses the electric current to rotate the shaft to a specific location in the rotation and stop. This can control the direction an item connected to the shaft points, such as moving a mechanical arm to a specific position.

Stepper motors use sophisticated controllers to apply voltage to the motor in steps to incrementally move the motor shaft through the rotation (thus the name stepper motor). The controller determines how many steps are required to position the shaft in the desired location. The locations are indicated by the number of degrees from a common point in the rotation.

Because the stepper motor requires a controller, you usually need to use additional software to communicate signals to the motor controller, telling it just where in the rotation cycle to place the motor shaft.

Servo Motors

The problem with stepper motors is that they can get out of sync with the controller. If something impedes the shaft's rotation, the controller doesn't realize the shaft isn't in the correct place. This can be bad for devices that require precise positioning, such as controllers for airplane wings.

Servo motors are a specialized application of stepper motors. They solve the positioning problem by adding a feedback system to the controller, so that the controller knows exactly where the motor shaft is pointing at all times. By using the feedback circuit, the controller can make on-the-fly adjustments to precisely position the shaft, even if the shaft is impeded along the way.

Servo motors are popular for use in devices that require fine control, such as robots and model airplanes or cars. Just as with stepper motors, servo motors require additional software to communicate signals to the motor controller. The Arduino library includes some libraries for working with servo motors. This allows you to use your Arduino for many types of high-precision motor applications.

Using DC Motors

When you use your Arduino to control a DC motor, you need to become familiar with three motor control aspects:

- Turning the motor on and off
- Controlling the motor speed
- Controlling the motor direction

This section goes through each of these three features of motors to demonstrate how you can use your Arduino to fully control any DC motor in your project.

Powering the Motor

Unfortunately, most motors require more voltage and current than what the Arduino can supply. That means you must connect the motor to an external power source to power the motor but somehow control that circuit using the Arduino interface. To do that, you need to use some type of relay that can separate the control signal from the switch.

You can use a variety of relays to control a motor circuit, both physical and electronic. For DC motors, the most popular solution is a transistor.

The transistor is a type of miniaturized relay. The transistor has three leads:

- The source, which is where you apply voltage to power the device
- The drain, which emits the voltage applied to the source
- The gate, which controls when the source and drain are connected

The gate lead behaves differently depending on the type of transistor. With an NPN-type transistor, when you apply power to the gate lead, voltage flows from the source lead to the drain lead of the transistor. When power is removed from the gate, the connection between the source and drain leads is broken and current stops flowing. With a PNP-type transistor, the opposite happens; current flows when there isn't a signal on the gate.

The trick to transistors is that you can use a very low voltage to control the gate but still use a larger voltage on the source. Thus, you can use a small voltage to control a larger voltage.

Figure 19.1 demonstrates using a transistor to control a motor circuit with your Arduino.

The key to selecting the right transistor to use is to ensure that the voltage and current rating can support the motor circuit. For most motor circuits that require extra current, the metal-oxide-semiconductor field-effect transistor (MOSFET) is the popular transistor to use. The MOSFET device is designed to handle higher voltages and currents, but still can be controlled by a small gate voltage. This is perfect for our Arduino projects.

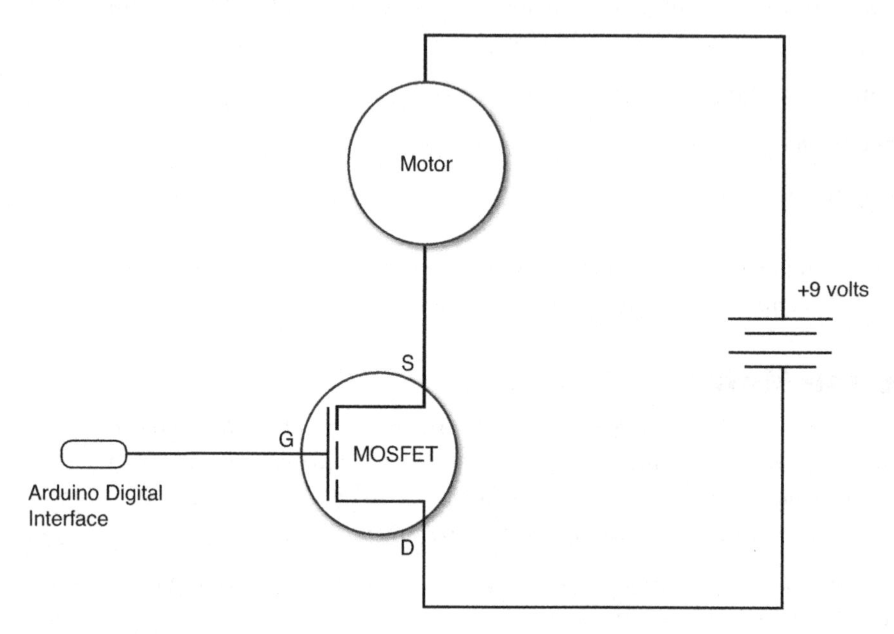

FIGURE 19.1
Using a transistor to control a motor circuit from your Arduino.

To control the transistor from your Arduino sketch, all you need to do is connect a digital interface to the gate on the transistor. When you set the digital interface to HIGH, the transistor will allow current to flow through the circuit to the motor. When you set the digital interface to LOW, the transistor will block current in the motor circuit.

BY THE WAY

Isolating the Arduino Pins

For very high-current motor circuits, it's recommended to place a resistor between the Arduino digital interface pin and the transistor gate lead. That way, if the transistor should short out, the Arduino is protected from the high current. For small DC motors, this isn't necessary.

Controlling Motor Speed

When the transistor is turned on by applying a voltage to the gate lead, the full power from the motor power source is applied to the motor, making it spin at a constant speed, based on the power source voltage.

If you remember from Hour 15, "Interfacing with Analog Devices," the Arduino supports a special type of digital output called a pulse-width modulation (PWM) signal. The PWM turns the digital output on and off at a predefined interval, called the duty cycle. You can use the PWM duty cycle to control the speed of your motor.

By applying a PWM to the gate of the transistor device, you can control how fast the transistor gate opens and closes, which in turn controls how fast voltage is applied to and removed from the motor. The on and off duty cycle happens so fast that the motor appears to be running at a slower constant speed. By simply changing the duty cycle of the PWM signal, you can change the speed of the motor.

In your Arduino sketches, you create a PWM output on a digital interface pin by using the `analogWrite` function:

```
analogWrite(pin, dutycycle);
```

A *`dutycycle`* value of 0 will stop the motor, and a *`dutycycle`* value of 255 will operate the motor at full speed. Any value in between will operate the motor at a slower speed.

Controlling Motor Direction

Using PWM solves the speed problem when working with motors, but it doesn't solve the direction problem. The motor turns in only one direction, depending on the polarity of the voltage applied to the motor leads. It would be somewhat cumbersome to physically change the wires going to a motor to get it to change directions.

There is, however, a fancy way to solve this problem, called an *H-bridge*. The H-bridge is a series of transistors interconnected so that the polarity of the voltage flow changes based on two control signals sent to the circuit. So, by using an H-bridge, you can control the motor direction by setting the digital state of the two control signals to determine which direction the motor operates, as shown in Table 19.1.

TABLE 19.1 H-Bridge Control Values

Control 1	Control 2	Result
LOW	LOW	The motor stops.
LOW	HIGH	The motor spins clockwise.
HIGH	LOW	The motor spins counterclockwise.
HIGH	HIGH	The motor stops.

The H-bridge circuit is extremely popular in motor use, and because of that, there are plenty of pre-built H-bridge circuits you can buy. Most H-bridges come in integrated circuit (IC) format, so they're easy to work with in your circuits. The Arduino Starter Kit includes an L293D H-bridge IC, which is ideal for most low-powered DC motors.

You control the H-bridge from your sketches by using three digital pins on the Arduino: one to turn the motor on and off, and two to control the motor direction. Your sketch just needs to keep track of which H-bridge control pin is connected to which digital interface pin.

Experimenting with Motors

Now that you've seen the basics of using DC motors with the Arduino, let's build some circuits to test them. This section walks through two exercises for working with DC motors.

Turning a Motor On and Off

For this experiment, you use a switch connected to the Arduino to control when a motor runs or stops. The hardware part of the circuit is a little involved, so let's go through that first. For the exercise, you need these parts:

- A low-current DC motor, found in most electronic supply stores
- A 9-volt battery clip with leads (along with a 9V battery)
- An NPN transistor, either a standard transistor (such as type TIP120) or a MOSFET transistor (such as type IRF510) will work for this project
- A general-purpose power diode
- A momentary contact switch
- A 1K-ohm resistor
- A 1-microfarad capacitor
- A breadboard

Figure 19.2 shows the circuit that you'll create for this exercise.

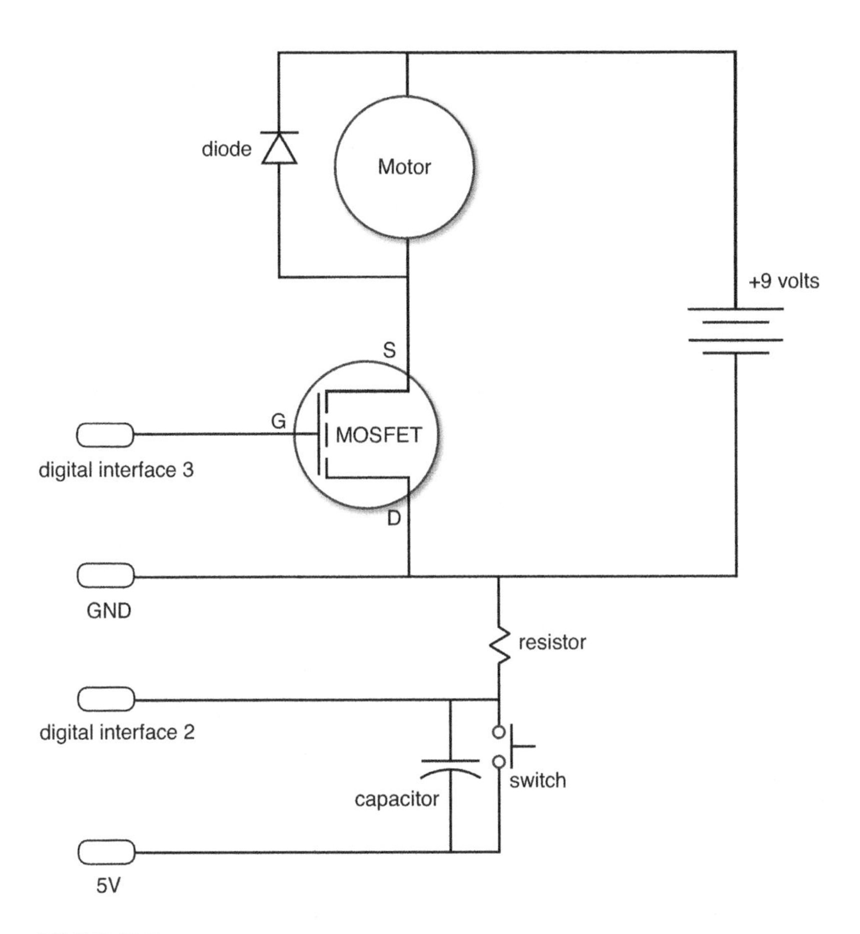

FIGURE 19.2
The DC motor circuit used for this exercise.

Here are the steps to follow to build the circuit:

1. Connect a jumper wire from the 5V pin on the Arduino to a rail on the breadboard.
2. Connect a jumper wire from the GND pin on the Arduino to another rail on the breadboard.
3. Connect the positive lead of the 9V battery clip to a third rail on the breadboard.
4. Connect the ground lead of the 9V battery clip to the same rail as the GND pin from the Arduino.

5. Place the momentary contact switch in the breadboard so that it straddles the middle of the breadboard and so that the contacts are connected to separate rails.
6. Connect the 1K-ohm resistor from one switch lead to the GND rail.
7. Connect the other switch lead to the 5V rail using a jumper wire.
8. Connect a jumper wire from the switch lead that's connected to the resistor to the Arduino digital interface 2 pin.
9. Connect the capacitor across both leads of the switch.
10. Place the transistor in the breadboard so that each lead connects to a separate rail and so that the flat side (or the side away from the heat sink on a MOSFET transistor) is on the left.
11. Connect a jumper wire from the top MOSFET transistor lead to the digital interface 3 pin on the Arduino. This is the gate lead. For standard transistors, the gate lead is the middle lead.
12. Connect the bottom transistor lead to the GND rail. This is the source lead, and is the same for both MOSFET and standard transistors.
13. Connect one lead from the motor to the middle lead of the MOSFET transistor. This is the drain lead. For standard transistors, this is the top lead.
14. Connect the other lead from the motor to the 9V rail on the breadboard (the positive pole of the battery clip).
15. Place the diode so that the cathode lead (the end marked with the line) is on the 9V rail on the breadboard and the anode lead connects to the middle pin of the MOSFET transistor, or the top lead of a standard transistor.

The motor runs from the nine-volt battery. The diode is used to prevent back voltage generated by the motor from entering the circuit and burning out the transistor or the battery. The capacitor is added to help prevent switch bounce.

After you've built the circuit, you're ready to start coding the Arduino sketch. Follow these steps:

1. Open the Arduino IDE, and enter this code into the editor window:

```
int switchPin = 2;
int motorPin = 3;
int led = 13;
int switchState = 0;

void setup() {
  pinMode(motorPin, OUTPUT);
  pinMode(led, OUTPUT);
```

```
  digitalWrite(motorPin, LOW);
  digitalWrite(led, LOW);
  attachInterrupt(0, changeMotor, RISING);
}

void loop() {
}

void changeMotor() {
  if (switchState == HIGH) {
    digitalWrite(motorPin, LOW);
    digitalWrite(led, LOW);
    switchState = LOW;
  } else {
    digitalWrite(motorPin, HIGH);
    digitalWrite(led, HIGH);
    switchState = HIGH;
  }
}
```

2. Save the sketch as **sketch1901**.
3. Click the Upload icon to verify, compile, and upload the sketch to your Arduino unit.

The sketch uses an external interrupt (see Hour 16, "Adding Interrupts") triggered by the switch to determine when power is applied to the transistor gate or not. I also used the pin 13 LED built in to the Arduino to indicate when the circuit should be on or off. That way you can troubleshoot things if the motor doesn't run.

Controlling Motor Speed

This exercise uses the exact same circuit as the previous exercise, but controls the transistor gate using a PWM generated on digital pin 3 of the Arduino. This enables you to control the speed of the motor by changing the duty cycle of the PWM output signal.

TRY IT YOURSELF 

Using PWM to Control a Motor

To control the motor speed this exercise uses the same external interrupt triggered by the switch to enable one of four speed settings for the motor. Each time you press the switch, the speed will change to a higher value, and then the motor will stop on the fourth switch press.

Here are the steps to create the sketch:

1. Open the Arduino IDE, and enter this code into the editor window:

```
int switchPin = 2;
int motorPin = 3;
int switchState = 0;

void setup() {
  analogWrite(motorPin, 0);
  pinMode(13, OUTPUT);
  digitalWrite(13, LOW);
  attachInterrupt(0, changeMotor, RISING);
}

void loop() {
}

void changeMotor() {
  switch(switchState) {
   case 0:
     analogWrite(motorPin, 75);
     digitalWrite(13, HIGH);
     break;
   case 1:
     analogWrite(motorPin, 150);
     digitalWrite(13, LOW);
     break;
   case 2:
     analogWrite(motorPin, 255);
     digitalWrite(13, HIGH);
     break;
  case 3:
     analogWrite(motorPin, 0);
     digitalWrite(13, LOW);
  }
  switchState++;
  if (switchState == 4)
     switchState = 0;
}
```

2. Save the sketch as **sketch1902**.
3. Click the Upload icon to verify, compile, and upload the sketch to your Arduino unit.
4. Press the switch once; the motor should start at a slow speed, and the pin 13 LED on the Arduino board should light.
5. Press the switch a second time; the motor should go a bit faster, and the pin 13 LED should go out.

6. Press the switch a third time; the motor should go at full speed, and the pin 13 LED should light.
7. Press the switch a fourth time; the motor should stop, and the pin 13 LED should go out.

Notice in the `changeMotor()` interrupt function that for each switch, press the duty cycle used in the `analogWrite()` function changes. This is what causes the motor to spin at a different speed. At duty cycle 0, the motor stops.

Using Servo Motors

Controlling servo motors is a bit trickier than working with DC motors. The servo motor controller requires specialized signals to position the motor shaft. Fortunately for us, the Arduino developers have created a prebuilt library that makes working with servo motors a breeze. This section discusses the Servo library, and then demonstrates how to use it in your Arduino sketches.

The Servo Library

The Servo library is installed by default in the Arduino IDE package. To use the functions in the Servo library, you must first create a `Servo` object variable in your sketch:

```
Servo myServer;
```

After you create the `Servo` object, you use the Servo library functions to connect the object to a pin on the Arduino and send signals to the servo motor. Table 19.2 lists the functions available inside the library.

TABLE 19.2 The Servo Library Functions

Function	Description
`attach(`*`pin`*`)`	Associates the `Servo` object to a digital pin.
`attached`	Returns a true value if the `Servo` object is attached to a digital pin.
`detach`	Removes the `Servo` object from any attached pins.
`read`	Returns the current position of the server, in degrees from 0 to 180.
`write(`*`deg`*`)`	Sets the servo shaft to the position specified by deg in degrees from 0 to 180.
`writeMicroseconds(`*`micro`*`)`	Sets the servo shaft to the position specified by micro microseconds.

Because servo motors operate as stepper motors, you can control the position of the motor shaft by either specifying a degree in the rotation (from 0 to 180 degrees) or as the number of microseconds the motor should step the shaft. For most applications, just specifying the degree value is sufficient; for fine-tuning to an exact location, however, you can use the microseconds value.

Typically, to work with the servo in your sketch, you just use three statements:

```
Servo myServo;
myServo.attach(5);
myServo.write(90);
```

The first statement creates the `Servo` object, the second statement attaches the object to digital interface pin 5, and then the third statement tells the servo to rotate the motor shaft to the 90-degree position.

Experimenting with Servos

This experiment allows you to control the position of a servo motor using a potentiometer. As you change the potentiometer position, the shaft of the servo motor turns to a new position.

▼ TRY IT YOURSELF

Positioning a Servo Motor

The Servo library in the Arduino IDE provides functions to control the position of a servo motor by specifying a degree value. The servo motor can position the motor shaft along a 180-degree arc.

Fortunately, the circuit for this exercise isn't as complex as the previous one. All you need for this circuit are the following:

- A low-power servo motor
- A 10K-ohm potentiometer
- An 100-microfarad electrolytic capacitor
- A breadboard

Most hobbyist servo motors are low powered, so you should be able to find one that operates within the 5V power of the Arduino at any electronic parts distributor. Also, because the servo motor runs on 5 volts, you don't have to worry about using an external power source.

This exercise also uses a potentiometer to control the servo motor position. You'll link the analog input from the potentiometer to the server output so rotating the potentiometer shaft rotates the servo motor. The electrolytic capacitor is used to help prevent any voltage spikes that may occur when the servo motor starts from damaging the potentiometer or the Arduino.

Here are the steps for building the circuit required for the exercise:

1. Connect the 5V pin on the Arduino to a rail on the breadboard using a jumper wire.
2. Connect the GND pin on the Arduino to a rail on the breadboard using a jumper wire.
3. Connect three jumper wires to the servo motor connector.
4. Connect the jumper wire from the black servo motor lead to the GND rail on the breadboard.
5. Connect the jumper wire from the red servo motor lead to the 5V rail on the breadboard.
6. Connect the jumper wire from the white servo motor lead to the digital interface 5 pin on the Arduino. This is the servo control interface.
7. Place the capacitor so that the positive lead is connected to the 5V rail and so that the other lead is connected to the GND rail.
8. Place the potentiometer on the breadboard so that the three leads are connected to separate rails.
9. Connect one outer lead from the potentiometer to the 5V rail on the breadboard.
10. Connect the other outer lead from the potentiometer to the GND rail on the breadboard.
11. Connect the middle lead from the potentiometer to analog interface 0 on the Arduino.

That completes the circuit. Figure 19.3 shows a diagram of the completed circuit.

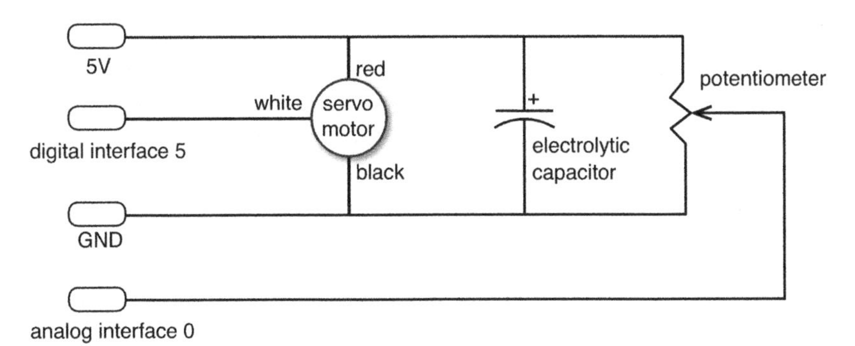

FIGURE 19.3
The servo motor circuit.

Now for coding the sketch. This exercise uses the Servo library that's installed in the Arduino IDE by default, so follow these steps to use it:

1. Open the Arduino IDE, select Sketch from the menu bar, and then select Import Library, Servo.
2. Enter this code in the editor window:

```
#include <Servo.h>

Servo myServo;
int position;

void setup() {
  myServo.attach(5);
}

void loop() {
   position = analogRead(0);
   position = map(position, 0, 1023, 179, 0);
   myServo.write(position);
   delay(10);
}
```

3. Save the sketch as **sketch1903**.
4. Click the Upload icon to verify, compile, and upload the sketch to your Arduino unit.
5. When the sketch starts to run, slowly turn the potentiometer shaft. The servo motor should move as you turn the shaft.

The sketch uses the `map` function to map the 1024-value range generated by the `analogRead` function into a 180-value range used to position the servo motor. Notice that I had to reverse the mapping to get my Arduino to move the servo shaft in the same direction that I moved the potentiometer shaft. Depending on how you connect your potentiometer, you may have to reverse that mapping in your setup.

BY THE WAY

The Motor Shield

If you do a lot of work with motors, you may be interested in exploring the Arduino Motor Shield. The Motor Shield provides a standard interface for connecting up to two DC brushed motors and one servo motor. It also provides feedback on the current and voltage used by the motors in the circuit.

Summary

This hour explored the world of motors. Many applications require motors to move mechanical parts, from building robots to automating signs. You can use your Arduino to control both simple DC motors as well as more complicated stepping and servo motors. You can control the speed of a DC motor by using the PWM feature available on some of the digital interface ports. To control the direction of a motor, you need to use an external H-bridge and control it with two digital interface ports. For using servo motors, you need to use the Servo library, installed by default in the Arduino IDE. You can control the exact location of the servo motor shaft using simply commands within the Servo library.

In the next hour, you'll learn another popular use of the Arduino: controlling LCD displays. Often, you need a quick interface to see the status or change the settings in a sketch, and connecting your Arduino to a computer using the serial monitor may not be an option. By adding a simple LCD to your Arduino, you can easily display short messages.

Workshop

Quiz

1. What type of motor should you use to precisely position a model airplane wing without error?

    A. Servo motor

    B. Stepper motor

    C. DC motor

    D. AC motor

2. Can you reverse the direction of a DC motor without having to physically reconnect it to the circuit? Yes or no.

3. What type of signal do you need to use to control the speed of a motor?

Answers

1. A. The servo motor includes circuitry to precisely control the location of the motor shaft.

2. Yes, you can use an H-bridge circuit to control the motor direction using two control signals.

3. By applying a pulse-width modulator (PWM) signal to the transistor gate, you can control the speed of the motor.

Q&A

Q. Can you control more than one servo motor from the same Arduino unit?

A. Yes, you can define multiple `Servo` objects in your sketch and attach each one to a separate digital interface pin.

Q. Does the Arduino contain a library for working with stepper motors?

A. Yes, you can use the Stepper library. It's included in the Arduino IDE package by default.

HOUR 20
Using an LCD

What You'll Learn in This Hour:

- The different types of LCD devices
- How to use an LCD with your Arduino
- How to use the LCD shield

So far in our experiments, we've used the serial monitor output on the Arduino to communicate information from our sketches. That's an easy way of communicating, but it does limit the use of the Arduino because you must have a computer connected to display the information. You can, however, display data from your Arduino in other ways without using a computer. One of the most popular methods is to use LCD devices. This hour demonstrates how to use LCD devices in your Arduino projects to output data from your sketches.

What Is an LCD?

Liquid crystal display (LCD) devices have been used by electronic devices for years to display simple alphanumeric information. The principle behind the LCD is to energize a series of crystals contained within a sealed enclosure to appear either opaque or transparent against a lighted background. The crystals are arranged in a pattern so that you can produce letters, numbers, and symbols based on which crystals are opaque and which ones are transparent.

Many different types of LCD devices are on the market. This section discusses some of the features that you need to be aware of as you look for LCD devices to use in your Arduino projects.

Display Types

You can use two basic types of LCD devices in your Arduino projects:

- Alphanumeric LCD devices
- Graphical LCD devices

An alphanumeric LCD device uses a small grid of lights to display letters, numbers, and simple symbols. The most common grid layout is a 5 × 8 grid of dots. The LCD displays each character by turning on or off each crystal in the grid to represent the character, as shown in Figure 20.1.

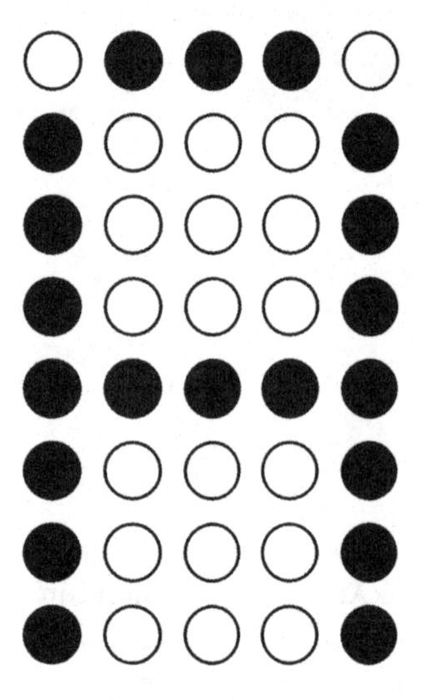

FIGURE 20.1
The 5 × 8 LCD grid layout displaying the letter *A*.

Alphanumeric LCD devices commonly include several character grids arranged in a line so that you can display words and sentences. The most common alphanumeric LCD devices used in the Arduino world is the 16 × 2 LCD device, which can display two lines of 16 characters.

Graphical LCD devices use a single larger grid of individual lights to display information. Instead of a separate grid for each character, graphical LCD devices provide a single array of crystal dots that you must control individually.

The most common graphical grid layout that you'll find for Arduino projects is the 128 × 64 LCD. The benefit of that layout is that you can display characters at any resolution you prefer; you're not limited to the 5 × 8 resolution used in alphanumeric LCD devices. Also, you can use the 128 × 64 layout as a canvas, creating complex drawings as well as numbers and letters.

Color Types

Besides the display type of the LCD device, you can also use different color patterns to display the characters. The LCD device uses two light sources. One light source is the color of the LCD crystals, and the other color is the background light that the crystals block. This produces two different ways to color the LCD:

- A negative LCD displays characters in color on a dark background.
- A positive LCD displays dark characters on a color background.

With the negative display type, you often have a choice of which color the characters appear in. However, you only have one color choice per LCD device.

With a positive LCD, small LEDs are used to light the background, so there are often more color choices. Another advantage of using a positive LCD is the RGB type of background; it provides three background LEDs: red, green, and blue. You can adjust the intensity of each background light to produce just about any background color.

Interfacing with LCD Devices

As you might guess, there are lots of individual dots to turn on and off in an LCD device to produce output. Just trying to display a simple 16-letter sentence requires having to control 5 × 8 × 16 = 640 separate dots.

Fortunately for us, we don't have to worry about trying to turn individual dots on or off to display characters. Most LCD devices used in Arduino projects include a separate controller chip that interfaces with the LCD. That helps reduce the number of wires we need to use to control the LCD, and makes it easier to send signals to display characters. This section discusses how to use these types of LCD devices with your Arduino.

LCD Device Wiring

The most popular LCD devices that you'll find for Arduino projects use the HD44780 controller chip to manage the LCD. That chip uses a series of 16 interface pins, shown in Table 20.1.

TABLE 20.1 The HD44780 Interface Pins

Pin	Name	Description
1	VSS	Ground connection
2	VDD	+5V connection
3	VO	Contrast adjustment
4	RS	Register selection
5	RW	Read/write
6	EN	Enable
7	DO	Data line 0
8	D1	Data line 1

Pin	Name	Description
9	D2	Data line 2
10	D3	Data line 3
11	D4	Data line 4
12	D5	Data line 3
13	D6	Data line 6
14	D7	Data line 7
15	A	Backlight anode
16	K	Backlight cathode

The 16 interface pins are usually located in the upper-left side of the LCD device. Figure 20.2 shows an example of an LCD device that uses an HD44780 controller chip.

FIGURE 20.2
A monochrome LCD device using the HD44780 chip.

The LCD uses long header pins that easily plug into a breadboard, or that can be soldered into a printed circuit board. All you need to do is connect the LCD pins to your Arduino digital interface pins.

However, you don't need to dedicate 16 pins on your Arduino to communicate with the HD44780 chip. The really neat thing about the HD44780 LCD controller chip is that it can operate in two modes: 8-bit mode or 4-bit mode.

In 4-bit mode, you need to use only four data lines to send the character data, which saves on the number of wires you need to interface from your Arduino to the chip. All you need is six wires—data lines 4 through 7, the EN line, and the RS line.

BY THE WAY

Multicolor Backlighting

The LCD kits that support multicolor backlights have 18 interface pins rather than 16. Pins 16, 17, and 18 control the red, green, and blue LEDs for the backlight color. You can control the LCD background color by sending pulse-width modulation (PWM) signals to each of those three pins.

Connecting the LCD to Your Arduino

One of the more complicated parts of using an LCD device is getting it wired to your Arduino. The first part of the process is deciding just what Arduino digital interfaces you have available to use to control the LCD. Remember, you need at least six digital interfaces, which decreases the number of interfaces you have available for working with sensors.

You must keep track of which digital interfaces you use for which LCD signal for your sketches. The easiest way to do that is to create a table that maps the digital interface ports you select to the LCD pins. Table 20.2 shows the map used for the examples in this hour.

TABLE 20.2 Mapping Arduino Interfaces to the LCD

Arduino Digital Interface	LCD Device Pin
2	4 (RS)
3	6 (EN)
4	11 (D4)
5	12 (D5)
6	13 (D6)
7	14 (D7)

Besides these pins, you also need to connect six more pins on the LCD device:

- Pin 1 connects to ground (GND).
- Pin 2 connects to +5 volts.
- Pin 3 connects to +5 volts, but through a potentiometer.
- Pin 5 connects to ground (GND).
- Pin 15 connects to +5 volts through a resistor.
- Pin 16 connects to ground.

Pins 1 and 2 provide power to the LCD device, and pins 15 and 16 power the LED backlight. Depending on the LCD device, you may or may not have to use a resistor to connect pin 15 to the +5 volts. Most LED backlights don't require the resistor, but some do. If in doubt, go ahead and use a small resistor, such as 220 ohms to help limit the current going to the LED backlight.

If you're using an RGB backlight LCD, you must connect pins 16, 17, and 18 to create the color background you want. You can connect these to PWM pins on the Arduino to vary the voltage applied to each; those signals control what color appears in the background.

Pin 3 controls the contrast of the LCD characters. You can place a potentiometer between pin 3 and the +5V so that you can adjust how bright the display appears. The size of the potentiometer doesn't matter, but the larger the value the less sensitive the contrast control will be.

Figure 20.3 shows the completed schematic for wiring the LCD to your Arduino.

After you've mapped out what pins you need to connect, place the LCD connectors in a breadboard, and use jumper wires to connect them to the proper place. It helps to connect the Arduino +5 and GND pins to rails on the breadboard to make it easier to connect the various LCD pins that require power or ground.

Once you have the LCD device wired to your Arduino, you're ready to start programming your sketches.

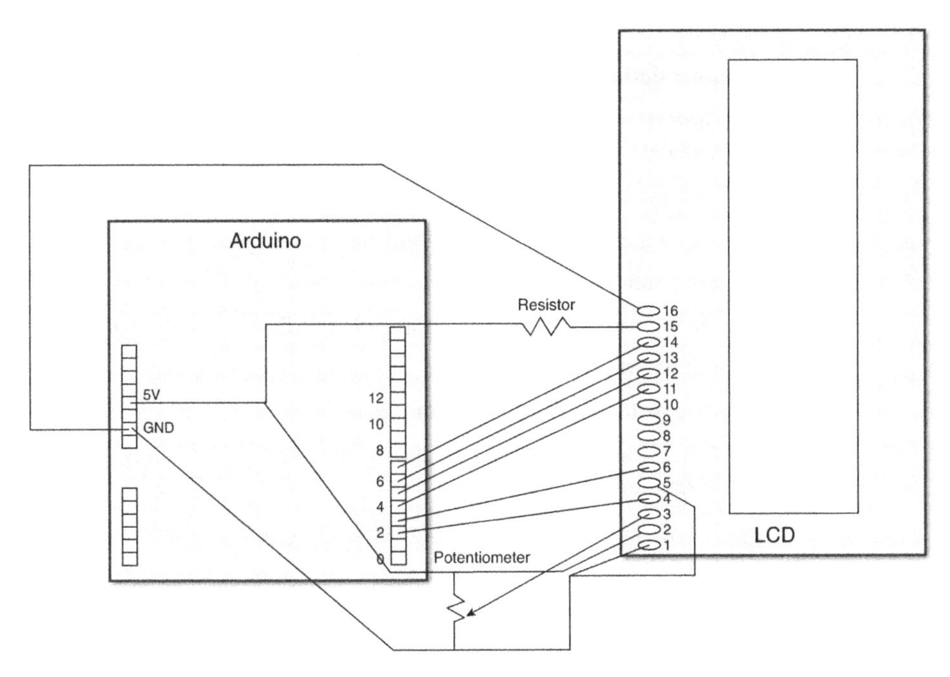

FIGURE 20.3
The Arduino LCD wire schematic.

The LiquidCrystal Library

By now, you should expect that the resourceful developers that are part of the Arduino community would have created a custom library for working with LCD devices—and you'd be correct! The LiquidCrystal library, which is installed by default in the Arduino IDE package, makes it a snap to interface with LCD devices from your sketches.

The LiquidCrystal Functions

The LiquidCrystal library defines a series of functions that you use to output data to the LCD. Table 20.3 describes these functions.

TABLE 20.3 The LiquidCrystal Library Functions

Function	Description
`LiquidCrystal(`*`rs, en, d4, d5, d6, d7`*`)`	Creates an LCD object and defines the LCD interface pins on the Arduino.
`autoscroll`	Enables automatic scrolling.
`begin(`*`cols, rows`*`)`	Defines the number of rows and columns in the LCD.
`blink`	Continually blinks the cursor.
`clear`	Erases the LCD and places the cursor in the upper-left corner.
`createChar`	Defines a special character that you can use in the output.
`cursor`	Display the cursor as an underscore.
`display`	Turn on the LCD device, displaying any data currently in the output buffer.
`home`	Places the cursor in the upper-left corner, without erasing the output.
`leftToRight`	Sets the direction new characters display starting at the left and going toward the right.
`noAutoscroll`	Disables automatic scrolling.
`noBlink`	Disables the blinking cursor.
`noCursor`	Disables the cursor display.
`noDisplay`	Blanks the display output, but retains the output data.
`print(`*`data, BASE`*`)`	Prints text to the LCD device. For numeric values, you can specify BASE.
`rightToLeft`	Sets the direction new characters display starting at the right and going toward the left.
`scrollDisplayLeft`	Scrolls the display contents one position to the left.
`scrollDisplayRight`	Scrolls the display contents one position to the right.
`setCursor(`*`col, row`*`)`	Places the cursor at a specific row and column location.
`write(`*`char`*`)`	Sends a single character to the LCD device.

The `LiquidCrystal` function is a little odd in that it acts like an object-oriented object rather than a function. You define an object of the `LiquidCrystal` type, and then you can use the other library functions on that object. The next section walks through how to do that.

Using the Library

To use the LiquidCrystal library in your Arduino shield, you first must define a `LiquidCrystal` object:

```
LiquidCrystal lcd(2, 3, 4, 5, 6, 7);
```

The variable `lcd` becomes an object using the LiquidCrystal type. The parameters of the function define the Arduino digital interface pins that you connected to the LCD interface pins (RS, EN, D4, D5, D6, and D7). If you choose to use all eight data lines, the first four data lines can be specified first in the data line order (RS, EN, D0, D1, D2, D3, D4, D5, D6, D7).

After you create the `LiquidCrystal` object, you can use the library functions on that object:

```
lcd.begin(16, 2);
lcd.home();
lcd.print("This is a test");
lcd.setCursor(0, 2);
lcd.print("This is the end of the test");
```

The `begin` function defines the columns and rows available on the device. This sketch example uses a 16 × 2 LCD device to display two lines of 16 characters. The sketch uses the `setCursor` function to move the cursor to the second line in the LCD device before displaying the second line of text.

WATCH OUT!

Overflow

Notice in the second output line that I try to display more than 16 characters in the `print` function. On a 16 × 2 display, the output line is truncated after 16 characters. Some larger displays (such as 20 × 2) wrap the text to the next line in the device. Be careful that you know just how your specific LCD device operates before using it!

Using an LCD Device

Now that you've seen the basics of how to use an LCD device with your Arduino, let's go through an example.

▼ TRY IT YOURSELF

Displaying Data

This example uses an LCD device along with the LiquidCrystal library to display the temperature detected from a TMP36 sensor (see Hour 18, "Using Sensors").

First, connect the LCD device to your Arduino as described earlier in the "Connecting the LCD to the Arduino" section. Then, follow these steps to connect the temperature sensor:

1. Plug the TMP36 sensor into the breadboard so that the three leads are connected to three separate rails. Make sure the flat side of the sensor is facing toward your left.
2. Connect the top pin of the TMP36 sensor to the +5V rail on your breadboard.
3. Connect the bottom pin of the TMP36 sensor to the ground rail on your breadboard.
4. Connect the middle pin of the TMP36 sensor to the Analog 0 interface on the Arduino.

Now that you have your circuit ready, you can create the sketch. Here are the steps to do that:

1. Open the Arduino IDE, and then click Sketch, Import Library, and then select the LiquidCrystal library.
2. In the editor window, enter this code:

```
#include <LiquidCrystal.h>

int output;
float voltage;
float tempC;
float tempF;
LiquidCrystal lcd(2, 3, 4, 5, 6, 7);

void setup() {
   lcd.begin(16, 2);
   lcd.home();
   lcd.print("The temp is:");
}

void loop() {
   output = analogRead(A0);
   voltage = output * (5000.0 / 1024.0);
   tempC = (voltage - 500) / 10;
   tempF = (tempC * 9.0 / 5.0) + 32.0;
   lcd.setCursor(5, 2);
```

```
    lcd.print(int(tempF));
    delay(5000);
}
```

3. Save the sketch as **sketch2001**.

4. Click the Upload icon to verify, compile, and upload the sketch to your Arduino unit.

As soon as the sketch completes the upload process, your Arduino should display the temperature in the LCD device. If you have an alternative power source for your Arduino, disconnect the USB cable from the Arduino and plug in the alternative power source. Your Arduino should power on and then display the temperature. Try holding the TMP36 sensor to make the temperature rise, or try placing an ice cube in a plastic bag next to the sensor to make the temperature fall. The LCD display should display the updated temperature after the 5-second delay. The delay is necessary; otherwise, the LCD output would be continually changing.

BY THE WAY

Using the Contrast

Don't get too discouraged if you plug everything in and nothing appears on the LCD. Play around with the contrast potentiometer connected to pin 3 on the LCD device. You may have to turn it all the way to the end of the rotation before anything appears in the display.

The LCD Shield

Yet another help is the LCD shield for the Arduino. Created by the popular Adafruit electronics company, it combines a 16 × 2 LCD device with a series of buttons that plugs into the standard Arduino Uno shield format. Figure 20.4 shows the LCD shield.

The LCD shield includes six buttons. Four buttons along the left side of the shield are arranged to provide an up, down, left, and right interface for simple menu control. A fifth button is set to the side of those buttons to act as a selection button, and the sixth button is on the right side of the shield and interfaces with the Reset pin on the Arduino (because you can't reach the Arduino reset button with the shield installed). Also on the shield is a potentiometer for adjusting the brightness of the LCD.

The nice thing about the LCD shield is that instead of using six connections, it only uses three pins to interface with the Arduino. It does that by utilizing the I2C interface (see Hour 17, "Communicating with Devices").

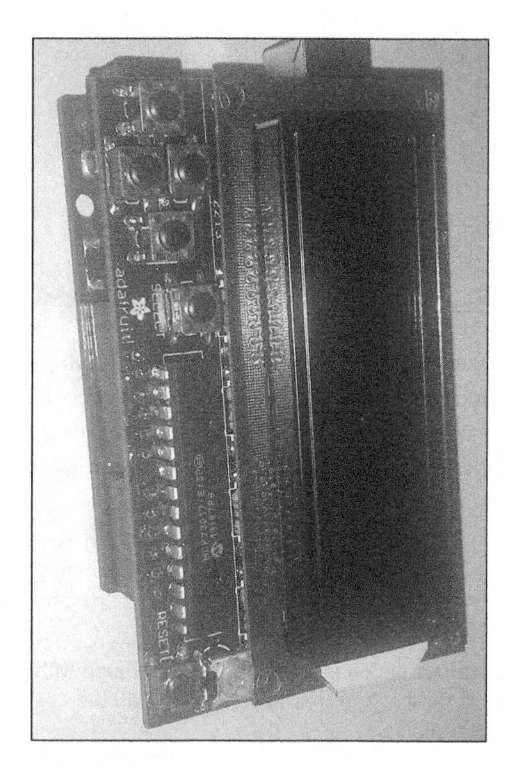

FIGURE 20.4
The Arduino LCD shield.

The Arduino sends data using the I2C protocol to the chip on the LCD shield, which decodes the signals and converts them to drive the HD44780 controller chip.

This section walks through installing the LCD shield library and using it in a sketch to display data on the LCD shield.

Downloading and Installing the Library

The LCD shield comes complete with its own library. The LCD shield library replicates all of the features of the standard LiquidCrystal library, plus adds a couple of customized functions specific to the shield.

To download and install the LCD shield library into your Arduino IDE environment, follow these steps:

1. Read the instructions for downloading and installing the LCD shield library on the Adafruit website:

   `https://learn.adafruit.com/rgb-lcd-shield/using-the-rgb-lcd-shield`

2. Click the link to download the Adafruit RGB LCD shield library. This will download the folder Adafruit-RGB-LCD-Shield-Library-master.

3. Copy the Adafruit-RGB-LCD-Shield-Library-master folder to your Arduino libraries folder, usually located under the Documents folder for your Windows or OS X user account.

4. Rename the folder to **LCDShieldLibrary**.

Now when you look in the Import Library feature in the Arduino IDE, you should see the LCDShieldLibrary listed in the Contributed section. If so, you're ready to start using it.

The LCD Shield Library Functions

The LCD shield library uses the same format as the LiquidCrystal library, including all the same function names. That makes migrating your application from a standard LCD device to the LCD shield a breeze.

The only difference is in the initial object that you create. Instead of a `LiquidCrystal` object, you use the following:

```
Adafruit_RGBLCDShield lcd = Adafruit_RGBLCDShield();
```

After you define the object, you can use the same features as the standard LiquidCrystal library:

```
lcd.begin(16,2);
lcd.setCursor(0,1);
lcd.print("This is a test");
```

The LCD shield library adds two new functions of its own:

- **`setBacklightColor`:** Sets the background color of RGB LCD devices using standard color names.
- **`readButtons`:** Retrieves the status of all the buttons on the LCD shield.

The `setBacklightColor` function makes it easy to use RGB backlit LCD devices. You don't have to worry about setting pin voltages on the LCD device; all you do is specify what color you want to use for the background in the `setBacklightColor` function. Currently, it recognizes the colors red, yellow, green, teal, blue, violet, and white.

The `readButtons` function allows you to detect button presses on the LCD shield. You can tell which button was selected by using a logical AND operation on the output of the `readButton` function, with the labels `BUTTON_UP`, `BUTTON_DOWN`, `BUTTON_RIGHT`, `BUTTON_LEFT`, and `BUTTON_SELECT`:

```
button = lcd.readButtons();
if (button & BUTTON_UP) {
   lcd.print("The UP button");
else  if (button & BUTTON_DOWN) {
   lcd.print("The DOWN button");
}
```

BY THE WAY

Switch Bounce

The readButtons function in the LCD shield library uses code to eliminate switch bounce, so there's no need for you to add that yourself in your code.

Connecting the LCD Shield

The easiest way to use the LCD shield is to plug it directly into your Arduino. The pin layout of the LCD shield is designed to fit exactly into the standard Arduino Uno pin layout.

However, if you do that, you won't have access to any of the interface pins on the Arduino. You have two options to solve that problem:

- One option is to use a separate prototype board plugged between the Arduino and the LCD shield. Hour 24, "Prototyping Projects," shows how to do that. From the prototype shield, you can pull out the pins necessary to connect the sensor, such as the analog interface pin, and the +5V and GND pins for powering the sensor.
- The other method is to remotely connect the LCD shield board to the Arduino. Because the LCD shield uses the two I2C pins to communicate with the Arduino, you only need to connect four wires (the two I2C pins, the ground pin, and the +5V pin) between the LCD shield and the Arduino. Figure 20.5 demonstrates this.

The I2C pins are located in different places on the different Arduino models. The Uno uses analog interface 4 and 5, the Mega uses digital interface pins 20 and 21, and the Leonardo uses digital interface pins 2 and 3. The nice thing about using an Arduino Uno is that all the pins you need are in the bottom interface of the LCD shield. That way you don't have to plug the entire shield into your breadboard.

WATCH OUT!

The Ground Pin

Be careful when connecting the GND pin on the LCD shield device. The GND pin next to the 5V pin is not connected; you must use the GND pin that's next to the Vin pin.

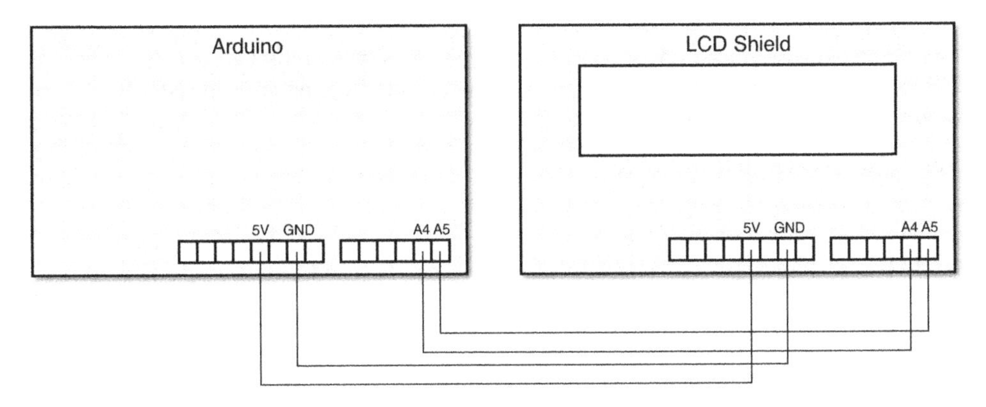

FIGURE 20.5
Connecting the LCD shield remotely to the Arduino Uno.

After you connect the LCD shield, you're ready to start coding. The next section walks through an updated temperature sensor example.

Using the LCD Shield

Let's update the temperature sensor example that we worked on earlier using the LCD shield. Make sure that you have your LCD shield connected as shown in the "Connecting the LCD Shield" section, and then work through this example.

TRY IT YOURSELF ▼

Displaying Data Using the LCD Shield

In this example, not only do you use the LCD shield to display the temperature, but you also use two of the buttons on the shield to toggle between displaying the temperature in Celsius and Fahrenheit.

You'll use the same TMP36 sensor, connected to analog interface 0, as you did before.

To create the sketch for the example, follow these steps:

1. Open the Arduino IDE, and select the Sketch menu option to import the LCDShieldLibrary library.
2. In the editor window, add a line before the two `#include` lines:

```
#include <Wire.h>
```

The LCD shield library uses some special data types that require the Wire.h library, but unfortunately at the time of this writing, the library doesn't automatically include that library file, so you need to manually type it into the code.

3. In the editor window, add this code:

```
#include <Wire.h>
#include <Adafruit_MCP23017.h>
#include <Adafruit_RGBLCDShield.h>

int output;
char temp[] = "C";
float voltage;
float tempC;
float tempF;
Adafruit_RGBLCDShield lcd = Adafruit_RGBLCDShield();

void setup() {
   lcd.begin(16, 2);
   lcd.home();
   lcd.print("The temp is:");
}

void loop() {
   output = analogRead(A0);
   voltage = output * (5000.0 / 1024.0);
   tempC = (voltage - 500) / 10;
   tempF = (tempC * 9.0 / 5.0) + 32.0;
   lcd.setCursor(5, 1);
   uint8_t button = lcd.readButtons();
   if (button & BUTTON_UP) {
     strcpy(temp, "C");
   } else if (button & BUTTON_DOWN) {
      strcpy(temp, "F");
   }
   if (strcmp(temp, "C")) {
      lcd.print(int(tempC));
      lcd.print(" C");
   } else {
      lcd.print(int(tempF));
      lcd.print(" F");
   }
}
```

4. Save the file as **sketch2002**.

5. Click the Upload icon to verify, compile, and upload the sketch to your Arduino unit.

When the sketch starts, it should show the temperature in Fahrenheit. Press the Down button on the LCD shield, and the output should change to show the temperature in Celsius. Press the Up button to change the temperature display to Fahrenheit.

Summary

This hour showed you how to use LCD devices in your Arduino projects. You can use a plain LCD device by connecting six wires from the Arduino to the LCD. The EN and RS lines must be connected to digital interfaces on the Arduino, along with four data lines D4, D5, D6, and D7. The LiquidCrystal library, which is installed by default in the Arduino IDE, provides an easy way to send data to the LCD device to display.

The hour also showed you how to use the popular LCD shield created by Adafruit. The LCD shield can either plug directly into the Arduino Uno interfaces or you can remotely connect it using a breadboard and jumper wires. You only need to connect four wires: the two I2C pins, the +5 pin, and the GND pin.

The next hour explores another popular shield used in Arduino projects: the Ethernet shield. The Ethernet shield provides an easy way to connect your Arduino to a network to both send and receive data with other network devices.

Workshop

Quiz

1. Which Arduino library do you use to interface with an LCD device?
   - **A.** The Wire library
   - **B.** The EEPROM library
   - **C.** The LiquidCrystal library
   - **D.** The SPI library
2. The LCD shield uses six wires to interface with the Arduino. True or false?
3. What LiquidCrystal library function do you use to move the cursor to the second line in the LCD device to display more data?

Answers

1. C. The Arduino IDE includes the LiquidCrystal library by default, which allows you to easily interface with standard LCD devices.

2. False. The LCD shield requires only two I2C connections (the +5V connection and a GND connection), totaling four wires.

3. The `setCursor` function allows us to specify the row and column location of the cursor. The output from the print or write functions will appear at the location of the cursor in the LCD device.

Q&A

Q. Can I use more than one LCD device on an Arduino at the same time?

A. Yes, it's been done! Because you only need six connections to the LCD device, you can connect two separate LCD devices directly to your Arduino and create two separate LiquidCrystal objects, each one pointing to the appropriate digital interface lines.

Another option that some developers have used, though, is to share the four data lines and the RS line with multiple LCD devices. Each device then connects to a separate digital interface for the EN signal. The LCD device only reads the data lines when the EN signal is HIGH, so you can control which device receives the output data by controlling which EN signal is set HIGH.

HOUR 21
Working with the Ethernet Shield

What You'll Learn in This Hour:

- How to connect your Arduino to the network
- How to connect to remote devices
- How to allow other devices to connect to your Arduino

These days, just about everything is connected to a network. Using networks makes it easy to query data from a centralized server and to provide data to multiple clients. Being able to connect your Arduino to a network can open a whole new world of possibilities with monitoring sensors. You can connect your Arduino to Ethernet networks in a few different ways. This hour discusses these different ways and walks through an example of providing sensor data to network clients.

Connecting the Arduino to a Network

You can connect your Arduino to Ethernet networks a few different ways. If you already have an Arduino unit, you can use a couple of shields to easily add network capabilities. If you're looking to purchase a new Arduino, there's an Arduino model that includes a wired Ethernet connection built in. This section discusses the different options you have available for using your Arduino on an Ethernet network.

The Ethernet Shield

If you already have an Arduino device that uses the standard Arduino pin layout, you can use the Ethernet Shield to connect to a wired network. The Ethernet Shield provides an RJ-45 jack for Ethernet connections, and also includes an SD card reader to provide extra space for storing data. Figure 21.1 shows the Ethernet Shield plugged into an Arduino Uno device.

FIGURE 21.1
The Ethernet Shield.

The Ethernet Shield uses the Serial Peripheral Interface (SPI) communication protocol (see Hour 17, "Communicating with Devices") to transfer data to the Arduino unit. The SPI interface uses digital interface pins 10, 11, 12, and 13 on the Arduino Uno, and pins 10, 50, 51, and 52 on the Arduino Mega unit.

A great feature of the Arduino Ethernet Shield is that it also supports the Power over Ethernet (PoE) feature. This feature allows the Ethernet Shield to get its power from network switches that support PoE. Therefore, if you have a PoE-capable network switch, you can power your Arduino and the Ethernet Shield directly from the network cable, without a separate power source.

Besides the PoE feature, the Ethernet Shield also includes pass-through header pins, so you can piggy-back additional shields on top of the Ethernet Shield, and it also has separate transmit and receive LEDs to indicate when it receives and sends data on the network.

The WiFi Shield

The WiFi Shield is another way to provide access to Ethernet networks, using wireless network access points. With the growing popularity of inexpensive home wireless network access points, the WiFi Shield can provide easy network access for your Arduino from any place. It supports both the 802.11b and 802.11g network protocols.

The WiFi Shield also uses the standard Arduino shield pin layout, so it plugs directly into most Arduino models. As with the Ethernet Shield, the WiFi Shield communicates using the SPI protocol, and requires the same digital interface pins (10, 11, 12, and 13 on the Uno; and 10, 50, 51, and 52 on the Mega). However, besides these pins, the WiFi Shield also uses digital interface pin 7 to communicate a handshake signal with the Arduino.

The WiFi Shield uses four LED indicators:

- **Data (blue):** Shows when data is being received or transmitted.
- **Error (red):** Indicates a communication error on the network.
- **Link (green):** Shows when the network connection is active.
- **L9 (yellow):** Connected to digital interface pin 9.

You can control the L9 LED as a status indicator from your sketch. This makes for a handy replacement of the pin 13 LED that's on the Arduino board, because you won't be able to see it with the shield plugged in.

The Arduino Ethernet Model

One Arduino model even has the Ethernet network built in to the device. The Arduino Ethernet model includes all the features of the Ethernet Shield (including an SD card reader), using an RJ-45 jack directly on the Arduino unit for the wired Ethernet connection.

The Arduino Ethernet unit mimics the layout of the standard Arduino Uno device. It includes 14 digital interface pins, but as with the Ethernet Shield, pins 10, 11, 12, and 13 are reserved for use with the Ethernet connection, leaving only nine digital interfaces that you can use in your sketches.

One downside to the Arduino Ethernet unit is that it doesn't include a USB connector to interface with your workstation. Instead, it uses a six-pin serial interface port that requires a USB to serial adapter, which you have to purchase separately.

The Ethernet Shield Library

All three Arduino networking options each have their own libraries for writing sketches that can communicate with the network. They all contain similar functions, so this hour does not cover all three libraries. Instead, this discussion focuses on the functions available in the Ethernet Shield library to show you the basics of using your Arduino on the network.

The Ethernet Shield library is installed in the Arduino integrated development environment (IDE) by default, and consists of five different classes:

- The `Ethernet` class, for initializing the network connection
- The `IPAddress` class, for setting local and remote IP addresses
- The `Client` class, for connecting to servers on the network
- The `Server` class, for listening for connections from clients
- The `EthernetUDP` class, for using connectionless User Datagram Protocol (UDP) messages with other devices on the network

Each class contains methods that help you interact with other network devices from the Ethernet Shield. The following sections show the methods contained in each class and how to use them.

WATCH OUT!

Referencing the Ethernet Library

If you select the Ethernet Library from the Arduino IDE Import Library feature, it will add `#include` directives for lots of different things, whether you use them in your sketch or not. This can make your sketch unnecessarily large. It's usually best to just manually include the `#include` directives required for your sketch. For the examples used in this hour, you'll only need the following:

```
#include <SPI.h>
#include <Ethernet.h>
```

The Ethernet Class

The Ethernet class contains three class methods:

- **`begin`:** Initializes the network connection.
- **`localIP`:** Retrieves the IP address of the Ethernet Shield.
- **`maintain`:** Requests a renewal of a Dynamic Host Control Protocol (DHCP) IP address.

Every sketch that uses the Ethernet Shield must use the Ethernet class `begin` method to initialize the shield on the network. The `begin` method has one required parameter and four optional parameters:

```
Ethernet.begin(mac, [ip], [dns], [gateway], [subnet]);
```

In Ethernet networks, every device must have a unique low-level address, called the Media Access Control (MAC) address. The MAC address is usually expressed as a 12-digit hexadecimal number, often using colons or dashes to separate the numbers in pairs to make it easier to read. For example, my Arduino Ethernet Shield MAC address is shown on the sticker as follows:

```
90-A2-DA-0E-98-34
```

Newer Arduino Ethernet Shield devices are each assigned a unique MAC address and have a sticker with the address on the device. Older Ethernet Shield devices aren't assigned a unique MAC address, so it's up to you to assign one with the `begin` method. Just make sure that the address you use doesn't match any other devices on your network.

Besides the MAC address, to interact on the network, your Arduino device will also need a unique IP address. The IP address identifies each network device within a specific subnetwork.

You can assign an IP address to a network device in two different ways:

- Manually assign a static address
- Request an address from the network dynamically

The following two sections discuss how to use both types of address methods with your Ethernet Shield.

Static Addresses

For the static address method, you're responsible for assigning a unique IP address to each device on your network. You do that by specifying both the MAC and IP addresses in the `begin` method:

```
Ethernet.begin(mac, ip);
```

You specify the MAC address using a byte array of hexadecimal values. For example:

```
byte mac[] = {0x90, 0xa2, 0xda, 0x0e, 0x98, 0x34 };
```

Unfortunately, you cannot specify the IP address as a string value. This is where the `IPAddress` class comes in. The `IPAddress` class allows you to specify an IP address using standard decimal numbers, but converts them into the byte format required for the `begin` method:

```
IPAddress ip(192, 168, 1, 10);
```

Besides the IP address, you may also need to assign a default router and subnet mask for your Arduino device to communicate outside of your local network, and may need to provide the IP address of the DNS server for your network:

```
IPAddress gateway(192, 168, 1, 1);
IPAddress dns(192, 168, 1, 1);
IPAddress subnet(255, 255, 255, 0);
```

After you define all the values, you just plug them into the `begin` method:

```
byte mac[] = {0x90, 0xa2, 0xda, 0x0e, 0x98, 0x34 };
IPAddress ip(192, 168, 1, 10);
IPAddress gateway(192, 168, 1, 1);
IPAddress dns(192, 168, 1, 1);
IPAddress subnet(255, 255, 255, 0);
Ethernet.begin(mac, ip, dns, gateway, subnet);
```

After you've assigned the IP address to the Ethernet Shield, you're ready to start communicating on the network.

Dynamic Addresses

The Dynamic Host Configuration Protocol (DHCP) provides a way for network devices to obtain an IP address automatically from a network DHCP server. Most wired and wireless home network routers include a DHCP server to assign IP addresses to devices on the home network.

To obtain an IP address for your Arduino using DHCP, you use the `begin` function, but only specify the MAC address for the device:

```
byte mac[] = {0x90, 0xa2, 0xda, 0x0e, 0x98, 0x34 };
Ethernet.begin(mac);
```

The Ethernet Shield will send a DHCP request onto the network, requesting an IP address (and any other network information such as the DNS server and default gateway) from the DHCP server.

When you use the `begin` method in DHCP mode, it will return a `true` value if a valid IP address is assigned to the Ethernet Shield, or a `false` value if not. You can test that in your code:

```
#include <SPI.h>
#include <Ethernet.h>

void setup() {
  Serial.begin(9600);
  byte mac[] = {0x90, 0xa2, 0xda, 0x0e, 0x98, 0x34 };
  if (Ethernet.begin(mac))
  {
     Serial.print("Found a valid IP address: ");
     Serial.println(Ethernet.localIP());
```

```
  } else
  {
      Serial.println("Unable to get an IP address");
  }
}

void loop() {
}
```

As shown here, you use the `localIP` method to retrieve the actual address assigned to the Ethernet Shield by the DHCP server. You'll need that whether you're using your Arduino as a server device, so that you know the address for your clients to connect to.

The `EthernetClient` Class

You use the `EthernetClient` class to connect to remote servers to exchange data. The client class allows you to send data to a remote host, as well as receive data from the remote host. The `EthernetClient` class has quite a few different methods, as shown in Table 21.1.

TABLE 21.1 The **`EthernetClient`** Class

Method	Description
`EthernetClient`	Creates a client object for communicating with a remote device using an Ethernet class object.
`connect(`*`address, port`*`)`	Connects to a remote host on the network.
`connected`	Checks whether the `EthernetClient` object is connected to a remote host. Returns `true` if connected, or `false` if not.
`available`	Returns the number of bytes available for reading.
`flush`	Removes any outbound data that hasn't been sent yet.
`read`	Reads a single byte from the connection.
`write(`*`val, len`*`)`	Sends 1 or more bytes of data to the remote host. If more than 1 byte, you must use an array and specify the length.
`print(`*`data, [BASE]`*`)`	Sends data to the remote host, using *`BASE`* (`DEC`, `HEX`, `OCT`) for printing numbers. Defaults to `DEC`.
`println(`*`data, [BASE]`*`)`	Sends data to the remote host, followed by the carriage return and newline characters.
`stop`	Disconnects from the remote host.

Because you need to start the connection only once, you usually connect to the remote server within the `setup` function in your sketch. Then in the `loop` function, you can loop between sending and receiving data, depending on what the server requires. The `connected` function allows you to check whether the remote host has disconnected the session, and the `available` function allows you to check if any data has been sent from the remote host.

A simple client connection to retrieve an FTP server welcome message from a remote FTP server looks like this:

```
#include <Ethernet.h>
#include <SPI.h>

byte mac[] = {0x90, 0xa2, 0xda, 0x0e, 0x98, 0x34 };
EthernetClient client;

void setup() {
   Serial.begin(9600);
   Ethernet.begin(mac);
   delay(1000);
   if (client.connect("ftp.ubuntu.com", 21))
   {
      Serial.println("Connected to server");
   } else
   {
      Serial.println("Connection attempt failed");
   }
}

void loop() {
   if (!client.connected())
   {
      Serial.println("Disconnected from remote host");
      client.stop();
      while(1);
   } else
   {
      if (client.available())
      {
         char incoming = client.read();
         Serial.print(incoming);
      }
   }
}
```

This code snippet uses the `EthernetClient` class to connect to the `ftp.ubuntu.com` FTP server on the standard FTP port (21). The FTP server returns a welcome message, which is received using the `read` method and displayed on the serial monitor. Figure 21.2 shows the output in the serial monitor after you run this program.

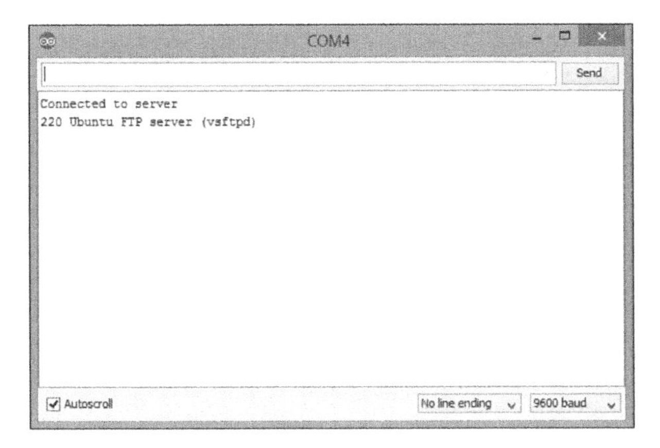

FIGURE 21.2
The serial monitor output showing the returned FTP server message.

BY THE WAY

Communicating with Servers

Network servers use standard protocols to communicate with clients. To successfully transfer data with a server, you need to know just what messages to send to the server and what messages to expect back. All approved standard Internet protocols (such as FTP and HTTP) are defined in Request For Comments (RFC) documents available on the Internet. It's always a good idea to study the appropriate RFC for the type of server you need to communicate with.

The `EthernetServer` Class

The key to building a network server is the ability to listen for incoming connection attempts from clients. With the Ethernet Shield library, you do that using the `EthernetServer` class. Table 21.2 shows the methods available in the `EthernetServer` class.

TABLE 21.2 The EthernetServer Class

Method	Description
`EthernetServer(port)`	Initializes a server port to listen for incoming connections.
`begin`	Starts listening for incoming connections.
`available`	Returns an `EthernetClient` object when a new connection is established with a remote client.
`write`	Sends data to all connected clients.
`print(data, [BASE])`	Sends data to all connected clients using the optional *BASE* if specified.
`println(data, [BASE])`	Sends data plus a newline character to all connected clients, using the optional *BASE* if specified.

To establish your Arduino as a network server device, you must complete the following four steps:

1. Use the `Ethernet` class to set an address on the network.
2. Use the `EthernetServer` class to create the server object.
3. Use the `begin` method of the `EthernetServer` class to listen for incoming connections.
4. Use the `EthernetServer available` method to accept an incoming client connection and assign it to an `EthernetClient` object.

It's common to use the `loop` function in an Arduino sketch to repeatedly check for new connections to the server object, as your Arduino can handle more than one client at a time. As each client connects to your server, you can assign it a new `EthernetClient` object.

You'll notice that the `EthernetServer` class only contains methods for writing data to the remote device; there aren't any for reading data from the remote device. The `EthernetServer` `write` and `print` methods are used for sending data to all clients connected to the Arduino. For sending data to a specific client, you use the `EthernetClient` class created for the connection:

```
byte mac[] = {0x90, 0xa2, 0xda, 0x0e, 0x98, 0x34 };
EthernetServer server(23);
void setup() {
   Serial.begin(9600);
   Ethernet.begin(mac);
   Serial.print("The server IP address is: ");
   Serial.println(Ethernet.localIP());
   server.begin();
}
```

```
void loop() {
   char incoming;
   EthernetClient client = server.available();
   if (client)
   {
      incoming = client.read();
      client.print(incoming);
   }
}
```

The `available` method in the `EthernetServer` class returns an `EthernetClient` object that represents the specific connection to the remote client. If no incoming client connections are available, it returns a `false` value.

The returned `EthernetClient` object persists in the sketch, so you can then use it to receive data from the client or send data to the client.

The `EthernetUDP` Class

You use the `EthernetServer` and `EthernetClient` classes in standard client/server environments where one device acts as a server, offering data, and another device acts as a client, retrieving data. This is standard for the Transmission Control Protocol (TCP) method of transferring data.

The UDP method of transferring data uses a connectionless protocol. No device acts as a server or client, all devices can send data to any other device, and any device can receive data from any other device.

The `EthernetUDP` class provides methods for interacting on the network as a UDP peer device (see Table 21.3).

TABLE 21.3 The `EthernetUDP` Methods

Method	Description
`begin(`*`port`*`)`	Initializes the library and assigns a local UDP port to listen on.
`write(`*`message`*`)`	Sends a character message to the remote host.
`read(`*`buf`*`, `*`size`*`)`	Reads up to *`size`* bytes of data sent from the remote host, placing it in *`buf`*.
`beginPacket(remoteIP,` *`remotePort`*`)`	Initializes a UDP session with the remote host using the remote port.
`endPacket`	Required after sending data in a remote UDP session.
`parsePacket`	Checks if a UDP packet is available for reading.
`available`	Returns the number of bytes available for reading.

Method	Description
stop	Releases all resources used for the UDP session.
remoteIP	Returns the IP address of the remote device.
remotePort	Returns the port address of the remote device.

Because UDP sessions are connectionless, sending and receiving data with a remote device is somewhat complicated. There isn't a connection established to control things, so your sketch must know when to send or receive data, and it must keep track of which remote device the data is going to or being read from.

Because of that, the `EthernetUDP` class requires a few different steps for both sending and receiving data. To send data to a remote device using UDP, you need to use the `beginPacket` method, then the `write` method, then the `endPacket` method, like this:

```
byte mac[] = {0x00, 0x00, 0x00, 0x00, 0x00, 0x00};
int localPort = 8000;
int remotePort = 8000;
IPAddress remoteIP(192, 168, 1, 100);
Ethernet.begin(mac)
EthernetUDP device;
device.begin(localPort);
device.beginPacket(remoteIP, remotePort);
device.write("Testing");
device.endPacket();
```

The `beginPacket` method specifies the IP address and port of the remote device. That's required for each data message you send, since a connection session isn't established between the two devices.

Similarly, to receive a UDP packet from a remote device, you must use this format:

```
byte mac[] = {0x00, 0x00, 0x00, 0x00, 0x00, 0x00};
int localPort = 8000;
char buffer[UDP_TX_PACKET_MAX_SIZE];
Ethernet.begin(mac)
EthernetUDP device;
device.begin(localPort);
int packetSize = device.parsePacket();
if (packetSize)
{
   IPAddress remoteIP = device.remoteIP();
   int remotePort = device.remotePort();
   device.read(buffer, UDP_TX_PACKET_MAX_SIZE);
}
```

The `EthernetUDP begin` method initializes the interface to receive UDP packets from any remote device on the network. The `parsePacket` method returns the number of bytes received from the remote device and ready to be read. You must use the `remoteIP` and `remotePort` methods to identify the specific remote device that sent the packet. If there is data to be read, you can then use the `read` method to retrieve the data to place in a buffer area.

Writing a Network Program

Let's walk through an example of using the Ethernet library to provide the data from a temperature sensor to remote clients on your network. We'll use a simplified network protocol instead of a standard Internet protocol. The Chat protocol listens for connections from clients, and then it echoes whatever data is sent to the server back to the client. Follow these steps to build your Chat server.

TRY IT YOURSELF ▼

Building a Chat Server

In this exercise, you use the Arduino Ethernet Shield, along with the Ethernet library to create a Chat server for your network. The Chat server will listen on the network for incoming client connections. When a client connects, the Chat server will echo whatever data the client sends to the server back to the client.

Here are the steps to create your Chat server:

1. Plug the Ethernet Shield into the header pins on your Arduino unit.
2. Plug the Arduino unit into the USB port on your workstation.
3. Open the Arduino IDE, and then in the editor window, enter this code:

```
#include <Ethernet.h>
#include <SPI.h>

byte mac[] = {0x90, 0xa2, 0xda, 0x0e, 0x98, 0x34 };
EthernetServer server(23);
void setup() {
   Serial.begin(9600);
   Ethernet.begin(mac);
   Serial.print("The server IP address is: ");
   Serial.println(Ethernet.localIP());
   server.begin();
}

void loop() {
   char incoming;
```

```
    EthernetClient client = server.available();
    if (client)
    {
      Serial.println("Client connected");
      while(client.connected()) {
        if (client.available()) {
          incoming = client.read();
          Serial.print(incoming);
          client.print(incoming);
        }
      }
      Serial.println("Client disconnected");
    }
}
```

4. Save the sketch as **sketch2101**.
5. Click the Upload icon to verify, compile, and upload the sketch to your Arduino unit.
6. Open the serial monitor tool in the Arduino IDE.
7. Press the Reset button on the Ethernet Shield. That button is connected to the Reset button on the Arduino and will restart your sketch.

When the sketch starts, you should see the IP address assigned to your Arduino appear in the serial monitor window. If not, make sure that the Ethernet cable is properly connected to the Ethernet Shield port, and to your network switch or router. Figure 21.3 shows the output from my Ethernet Shield on my network.

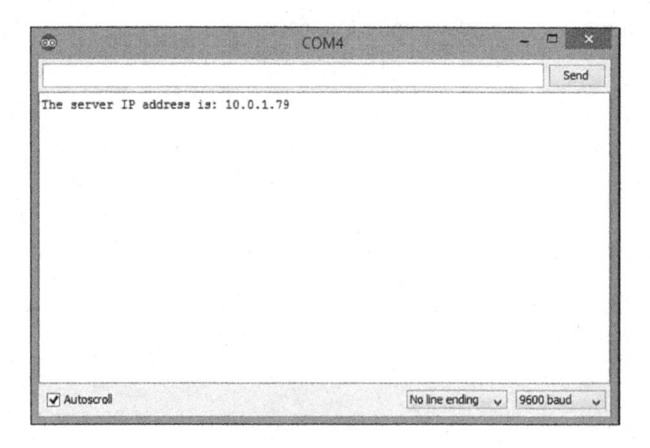

FIGURE 21.3
The output from the sketch2101 sketch, showing the IP address.

Next, you need to use a standard Telnet client software package to connect to your Arduino. For OS X and Linux users, that's easy; just open a Terminal window and enter the `telnet` command at the command prompt:

```
rich> telnet 10.0.1.79
```

For Windows users, you need to download and install a graphical Telnet client. The most popular one is the PuTTY package, at http://www.putty.org. After you've installed the PuTTY package, you can set a session to connect using Telnet to port 23, specifying the IP address assigned to your Arduino.

When you connect to the Chat server, you'll see some odd characters appear in the serial monitor. Those are binary characters used by the Telnet software to establish the connection. Once the connection is established, you can type characters in the client, and they will both appear in the serial monitor window, and echo back in the Telnet client window, as shown in Figure 21.4.

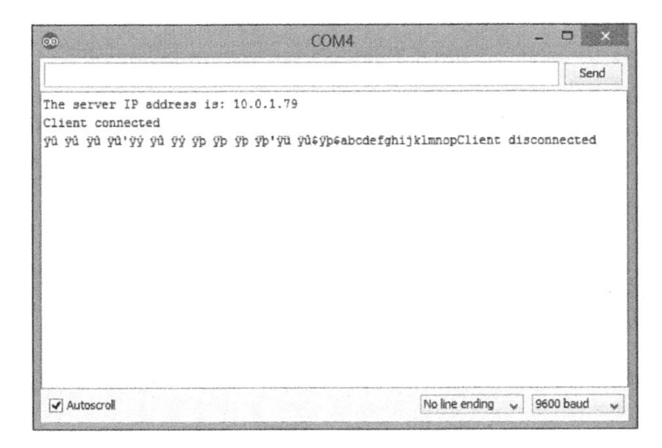

FIGURE 21.4
Output from the Chat server in the serial monitor.

You will most likely see two characters for each character that you type. One is displayed from the telnet client, and the other is what's returned by the chat server. Some telnet clients allow you to turn off the character echo if it knows the characters are being returned by the server.

Summary

This hour examined how to use the Arduino on Ethernet networks. Two shields are available to connect the Arduino to Ethernet networks. The Ethernet Shield connects to standard wired networks using an RJ-45 plug, and the WiFi Shield connects to wireless networks using either the

802.11b or 802.11g protocols. You can also use the Arduino Ethernet model, which includes the wired RJ-45 Ethernet connector built in.

The chapter then covered how to use the Ethernet library to write sketches that interface with other devices on the network. The `EthernetClient` class allows you to connect to remote devices, and the `EthernetServer` class allows you to listen for connections from clients. The hour finished by showing an example of how to write a simple server to communicate with clients on the network.

The next hour shows how to use your network capabilities to provide sensor data to others both on your local network and around the world!

Workshop

Quiz

1. What Ethernet method should you use to assign a static IP address to your Arduino unit?

   A. `localIP`

   B. `remoteIP`

   C. `begin`

   D. `connect`

2. The `EthernetUDP` class allows you to send connectionless packets to network devices without creating a session. True or false?

3. How do you receive data from a network client after initializing an `EthernetServer` object since there isn't a `read` method?

Answers

1. C. You use the `Ethernet.begin` method to define the MAC address of the Ethernet Shield, and you can also include a static IP address.

2. True. The EthernetUDP class contains methods for sending and receiving packets using UDP, which doesn't create a network session to the remote device.

3. The `EthernetServer.available` method returns an `EthernetClient` object. You then use the read method from that object to receive data from the remote client that connected to the server.

Q&A

Q. Can I place multiple Arduino units on my local network?

A. Yes, as long as you assign each Arduino a unique MAC address and a unique IP address.

Q. Can I use the LCD Shield with the Ethernet Shield?

A. Yes, but you may need to do some tweaking. When you plug the LCD Shield into the Ethernet Shield, the bottom of the LCD Shield may rest on the metal RJ-45 port on the Ethernet Shield, which can cause problems. You can purchase header pins to use as spacers between the two shields, providing enough separation for them to work. Because the LCD Shield uses Inter-Integrated Circuit (I2C) and the Ethernet Shield uses Serial Peripheral Interface (SPI), they won't conflict.

HOUR 22

Advanced Network Programming

What You'll Learn in This Hour:

- How to create a web server on your Arduino
- Retrieving sensor data with a web browser
- Controlling an Arduino from a web browser

In the previous hour, you saw how to use the Ethernet Shield to communicate on an Ethernet network from your Arduino device. That opens a whole new world of ways for you to communicate with your Arduino projects. This hour expands on that by showing how to provide sensor data to remote clients by using the Arduino as a web server and also how to control your Arduino from a remote client.

The Web Protocol

Thanks to the World Wide Web, the Hypertext Transfer Protocol (HTTP) has become the most popular method of transferring data on networks. Web browser client software comes standard on just about every workstation, tablet, and smartphone device. You can leverage that popularity with your Arduino programs by incorporating web technology to interface with your sketches.

Before you can do that, though, you need to know a little bit about how HTTP works. This first section walks through the basics of an HTTP session, and shows how to transfer data using HTTP servers and clients.

HTTP Sessions

HTTP uses a client/server model for transferring data. One device acts as the server, listening for requests from clients. Each client establishes a connection to the server and makes a request for data, usually a data file formatted using the Hypertext Markup Language (HTML). If the request is successful, the server sends the requested data back to the client and closes the connection. Figure 22.1 demonstrates this process.

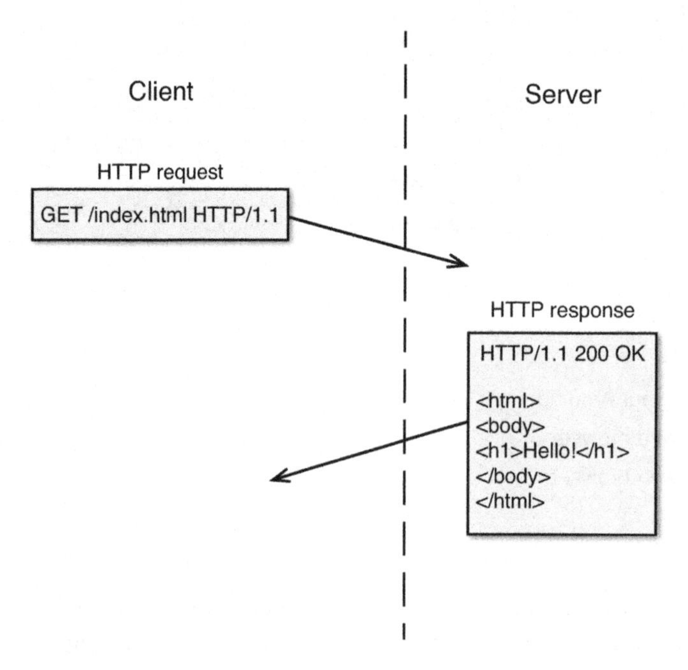

FIGURE 22.1
A typical HTTP session.

HTTP specifies how the client and server communicate with each other. The current standard for HTTP is version 1.1, and is defined by the World Wide Web Consortium (W3C). It specifies the exact format for each request and response message. The client makes a specially-formatted HTTP request to ask for the data, and the server responds with an HTTP response, along with the requested data.

The HTTP Request

The HTTP specifications define the client's request to the server. Because HTTP is a text-oriented protocol, the client HTTP request consists of all text, separated into three parts:

- A request line
- Request header lines (optional)
- An empty line terminated by a carriage return-line feed combination

The combination of the request line and header lines tell the server just what data the client wants to retrieve, along with some information about the client. Let's take a closer look at those parts.

The Request Line

The request line identifies the object the client is requesting from the server. It consists of three parts:

- A method token
- A Universal Resource Indicator (URI) identifying the requested data
- The protocol version

The method token defines the method action performed on the specified URI. Table 22.1 lists the valid method tokens currently supported in HTTP 1.1.

TABLE 22.1 HTTP Method Tokens

Token	Description
CONNECT	Converts the connection into a tunnel, often used for encrypted communication.
DELETE	Requests to delete the specified URI.
GET	Retrieve the specified URI.
HEAD	Requests the HTTP header of the response only.
POST	Sends data to the server.
PUT	Replaces the specified URI with the data provided.
OPTIONS	Retrieves the HTTP methods the server supports.
TRACE	Returns the request back to the client.

The `GET` and `POST` methods allow the client to send data to the server within the request. The `GET` method embeds the data with the URI itself, while the `POST` method sends the data as separate messages within the HTTP session.

The URI defines the path and filename of the data file you want to retrieve. The path is relative to the root folder of the web server.

The protocol version defines the HTTP version that the client is using.

A complete HTTP request looks something like this:

```
GET /index.html HTTP/1.1
```

This is the text the client sends to the web server to request the file `index.html` from the root folder of the web server.

Request Headers

After the request line, the client can optionally send one or more request headers. The request headers allow the client to send additional information about the request, about the connection, or even about the client itself (such as identifying the browser being used). The request header uses the following format:

```
header: value;
```

Each request header appears on a separate line and is terminated with a semicolon. The end of the request header list is indicated by an empty line with a carriage return and linefeed combination.

If your web client chooses to include request headers, a complete HTTP request would look something like this:

```
GET /index.html HTTP/1.1
User-Agent: Microsoft IE/10.0
Connection: Close
```

This request asks to retrieve the `index.html` file from the web server. It identifies the client web browser as the Internet Explorer package and tells the server to close the HTTP session connection after returning the response.

The HTTP Response

When the web server receives a request from a web client, it must formulate a response to return. The response consists of three parts:

- The status line
- One or more response header lines (optional)
- An empty line terminated by a carriage return-line feed combination

If the request is successful, the requested data follows immediately after the HTTP response. Just like the HTTP request, the HTTP response is a plain-text message. Let's take a look at the different parts of the response.

The Status Line

The status line returns the status of the request to tell the client how the server handled it. The status consists of three parts:

```
version status-code description
```

The version returns the HTTP version the server is using (usually HTTP/1.1). The status code and description indicate the status of the request. The status code is a three-digit code that indicates the status of the client's request. This allows the client to quickly identify success or failure of the request. If the request failed, the response code indicates a detailed reason why the request failed. There are five categories of HTTP status codes:

- **1xx:** Informational messages
- **2xx:** Success messages
- **3xx:** Redirection messages
- **4xx:** Client-side errors
- **5xx:** Server-side errors

Table 22.2 shows the full list of response status codes available in HTTP 1.1.

TABLE 22.2 HTTP Response Status Codes

Code	Text Description
100	Continue
101	Switching Protocols
200	OK
201	Created
202	Accepted
203	Non-Authoritative Information
204	No Content
205	Reset Content
206	Partial Content
300	Multiple Choices
301	Moved Permanently
302	Found
303	See Other
304	Not Modified
305	Use Proxy
307	Temporary Redirect
400	Bad Request
401	Unauthorized

Code	Text Description
402	Payment Required
403	Forbidden
404	Not Found
405	Method Not Allowed
406	Not Acceptable
407	Proxy Authentication Required
408	Request Time-Out
409	Conflict
410	Gone
411	Length Required
412	Precondition Failed
413	Request Entity Too Large
414	Request-URI Too Large
415	Unsupported Media Type
416	Requested Range Not Satisfiable
417	Expectation Failed
500	Internal Server Error
501	Not Implemented
502	Bad Gateway
503	Service Unavailable
504	Gateway Time-Out
505	HTTP Version Not Supported

The status-code numbers are always the same, but the text description may change depending on the server. Some servers provide more detail for failed requests.

The Response Header Lines

The response header lines allow the server to send additional information to the client besides the standard response code. The HTTP 1.1 version provides for lots of different response headers that you can use. Similar to request headers, response headers use the following format:

```
header: value;
```

Each response header is on a separate line and terminated with a semicolon. The end of the response header list is indicated by an empty line with a carriage return and linefeed combination.

A standard HTTP response would look something like this:

```
HTTP/1.1 200 OK
Host: myhost.com
Connection: Close
```

If any data is returned as part of the response, it should follow the closing carriage return and linefeed line of the header.

Now that you've seen how HTTP works, let's take a look at using it in an Arduino sketch to communicate with client workstations.

Reading Sensor Data from a Web Server

Currently, there isn't a standard Arduino library available for running a web server from your Arduino. Instead, you need to do a little coding to emulate the web server to remote clients. However, that's not as hard as you might think, thanks to the simplicity of HTTP and the HTML web page markup language.

This section walks through building a web server that returns the output from a temperature sensor using the Arduino Ethernet library to serve your sensor data on the network.

Building the Circuit

For the circuit, you need to connect a temperature sensor to your Arduino to provide the temperature information. In Hour 18, "Using Sensors," we worked with the TMP36 analog temperature sensor. The sensor provides an analog output signal that indicates the temperature. You just need to connect that output to an analog interface on the Arduino to process the data. Because the Ethernet Shield provides all the interface pins from the Arduino, you can plug your circuit directly into the Ethernet Shield interface header pins just as you would the regular Arduino interface header pins.

TRY IT YOURSELF ▼

Analog Temperature Sensor

To set up the analog temperature sensor for the web server project, follow these steps:

1. Place the TMP36 temperature sensor on the breadboard so that each of the three pins is in a separate rail row. Face the sensor so the flat edge is facing toward the left.

2. Plug the Ethernet Shield into the Arduino.
3. Connect the top pin of the TMP36 sensor to the +5 pin on the Ethernet Shield.
4. Connect the bottom pin of the TMP36 sensor to the GND pin on the Ethernet Shield.
5. Connect the middle pin of the TMP36 sensor to the analog interface 0 pin on the Ethernet Shield.
6. Plug a standard Ethernet network cable into the RJ-45 jack on the Ethernet Shield.
7. Plug the other end of the Ethernet network cable into your network switch or hub.

That's all the hardware you need for this exercise. Next comes writing the sketch.

Writing the Sketch

To build the server, you need to use the Ethernet library, which is installed by default in the Arduino IDE package. However, when you select the Ethernet library from the IDE interface, it adds more `#include` directives than you need for the sketch code, which will needlessly increase the size of your sketch when loaded on to the Arduino. Instead, we'll just manually add the `#include` directives to the code. Just follow these steps:

1. Open the Arduino IDE, and enter this code into the editor window:

```
#include <SPI.h>
#include <Ethernet.h>
byte mac[] = {0x90, 0xa2, 0xda, 0x0e, 0x98, 0x34};
EthernetServer server(80);

void setup() {
   Serial.begin(9600);
   Ethernet.begin(mac);
   delay(1000);
   Serial.print("The server is on IP address: ");
   Serial.println(Ethernet.localIP());
}

void loop() {
   EthernetClient client = server.available();
   if (client.connected()) {
      int temp = getTemp();
      client.println("HTTP/1.1 200 OK");
      client.println("Content-Type: text/html");
      client.println("Connection: close");
      client.println();
```

```
      client.println("<!DOCTYPE HTML>");
      client.println("<html>");
      client.println("<head>");
      client.println("<title>Current Temperature</title>");
      client.println("</head>");
      client.println("<body>");
      client.println("<h1>The Current Temperature</h1>");
      client.print("<h2>The current temperature is ");
      client.print(temp);
      client.println("&deg; F</h2>");
      if ((temp >= 68) && (temp <= 72))
         client.println("<p>It's a normal room temperature</p>");
      if (temp < 68)
         client.println("<p>It's a little cold in here!</p>");
      if (temp > 72)
         client.println("<p>It's a little warm in here!</p>");
      client.println("</body>");
      client.println("</html>");
      delay(10);
      client.stop();
   }
}

int getTemp() {
   int output;
   float voltage, tempC, tempF;
   output = analogRead(A0);
   voltage = output * (5000.0 / 1024.0);
   tempC = (voltage - 500) / 10;
   tempF = (tempC * 9.0 / 5.0) + 32.0;
   return int(tempF);
}
```

2. Save the sketch as **sketch2201**.

3. Click the Upload icon to verify, compile, and upload the sketch to your Arduino unit. (Make sure that you have the USB cable connected to your Arduino.)

4. Open the serial monitor tool from the Arduino IDE toolbar.

5. Press the Reset button on the Ethernet Shield. This resets the Arduino, restarting the sketch from the beginning. When the sketch starts, you should see the IP address assigned to your Arduino from your network appear in the serial monitor window.

Now your Arduino should be waiting for clients to connect to retrieve the temperature from the sensor. Just open a browser in a workstation on your network and enter the IP address of your Arduino (as shown in the serial monitor output). That should look something like this:

```
http://10.0.1.79/
```

You should get back a simple web page, showing the current temperature returned by the TMP36 sensor, as shown in Figure 22.2.

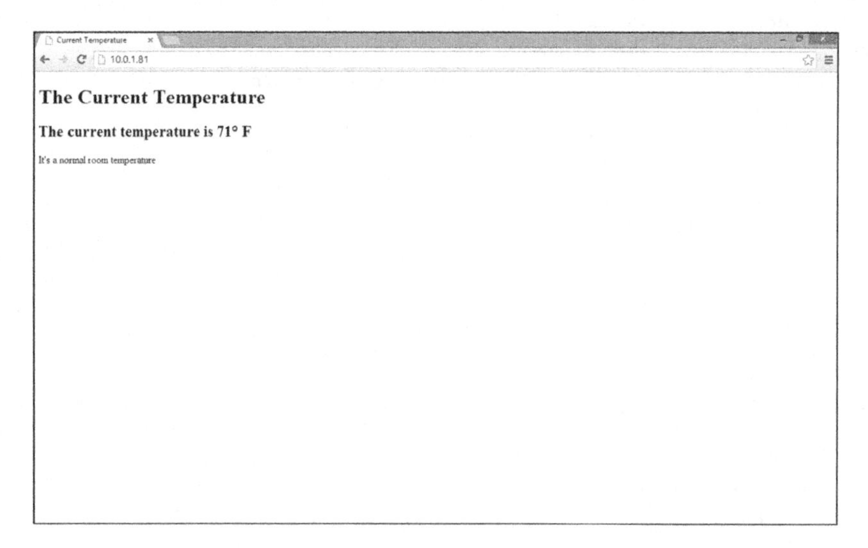

FIGURE 22.2
The web page generated by the sketch.

The simplicity of this example is in the web server code. We don't care what request the client sends to the web server, so the sketch doesn't bother trying to read the received data. We just assume that if a remote client is making a request to the web server, it wants to receive the current temperature back!

The first part of the data sent back to the client is the standard HTTP response. After that, the sketch sends an HTML-formatted document that contains the data from the temperature sensor.

Controlling an Arduino from the Web

The next step to using your Arduino on the network is the ability to control the Arduino outputs from a web client. For this exercise, you need to read the actual data sent by the web client, and then use that data to determine which output should be active or inactive.

For this exercise, you control a standard three-light traffic signal from a remote web client. When the web client connects to the Arduino, it will return a web page with three links: one for each light. The client can click a link to turn on the appropriate light.

Building the Circuit

First, you need to build the circuit. This section walks through what you need to set up the traffic signal for the sketch.

For this experiment, you need a few electronic components:

- Three LEDs (preferably red, yellow, and green, but they can be the same color if that's all you have available)
- Three 1K-ohm resistors (color code brown, black, red)
- A breadboard
- Four jumper wires

Once you gather these components, you can start the experiment.

TRY IT YOURSELF ▼

Digital Traffic Signal

For this experiment, you create a traffic signal that your Arduino will control using three separate digital interfaces. First, follow these steps to build the electronic circuit:

1. Place the three LEDs on the breadboard so that the short leads are all on the same side and so that the two leads straddle the space in the middle of the board so that they're not connected. Place them so that the red LED is at the top, the yellow LED in the middle, and the green LED is at the bottom of the row.
2. Connect a 1K-ohm resistor between the short lead on each LED to a common rail area on the breadboard.
3. Connect a jumper wire from the common rail area on the breadboard to the GND interface on the Arduino.
4. Connect a jumper wire from the green LED long lead to digital interface 5 on the Arduino.
5. Connect a jumper wire from the yellow LED long lead to digital interface 6 on the Arduino.
6. Connect a jumper wire from the red LED long lead to digital interface 7 on the Arduino.

That completes the hardware circuit. The circuit diagram for what you just created is shown in Figure 22.3.

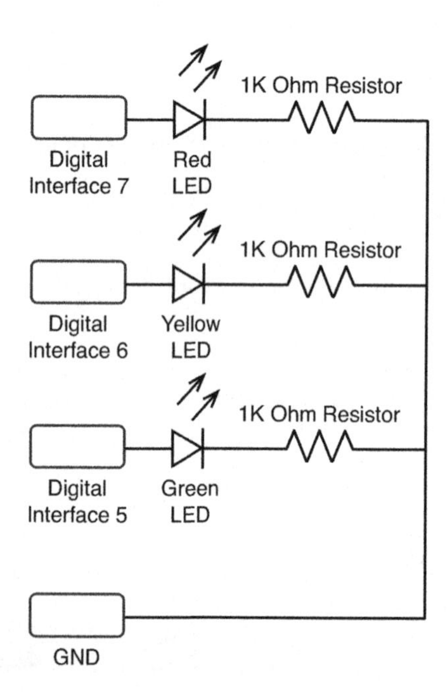

FIGURE 22.3
The circuit diagram for the traffic signal experiment.

Now you're ready to start coding the sketch that controls the traffic signal circuit from the Web.

Writing the Sketch

For the sketch, you need to create a web server that can read the request sent from a web client, parse the data sent, and activate or deactivate the appropriate LED.

▼ TRY IT YOURSELF

Creating the Web Server

For the web server sketch, you use the Ethernet library to listen for client connections and then read the data sent by the client. Unfortunately, you can only read the data 1 byte at a time. Because of that, the sketch will need to look for a specific character sequence to trigger which LED should light. The web server will expect the client to send the URI as follows:

```
http://<ipaddress>/?x
```

Where *<ipaddress>* is the IP address assigned to the Arduino, and *x* is the number of the LED you want to light:

- 1 for the red LED
- 2 for the yellow LED
- 3 for the green LED

The question mark before the number gives us a way to parse through the client data and know when to expect the control number to appear.

Here are the steps to create the sketch:

1. Open the Arduino IDE, and enter this code into the editor window:

```
#include <SPI.h>
#include <Ethernet.h>
byte mac[] = {0x90, 0xa2, 0xda, 0x0e, 0x98, 0x34 };
EthernetServer server(80);
int redLED = 7;
int yellowLED = 6;
int greenLED = 5;
char c;

void setup() {
   Serial.begin(9600);
   pinMode(redLED, OUTPUT);
   pinMode(yellowLED, OUTPUT);
   pinMode(greenLED, OUTPUT);
   digitalWrite(redLED, LOW);
   digitalWrite(yellowLED, LOW);
   digitalWrite(greenLED, LOW);
   Ethernet.begin(mac);
   delay(1000);
   Serial.print("The server is on IP address: ");
   Serial.println(Ethernet.localIP());
}

void loop() {
   EthernetClient client = server.available();
   if (client.connected()) {
     while (client.available()) {
       c = client.read();
       if (c == '?') {
         c = client.read();
         switch(c) {
           case '1':
```

```
            Serial.println("Activate Red LED");
            digitalWrite(redLED, HIGH);
            digitalWrite(yellowLED, LOW);
            digitalWrite(greenLED, LOW);
            break;
         case '2':
            Serial.println("Activate Yellow LED");
            digitalWrite(redLED, LOW);
            digitalWrite(yellowLED, HIGH);
            digitalWrite(greenLED, LOW);
            break;
         case '3':
            Serial.println("Activate Green LED");
            digitalWrite(redLED, LOW);
            digitalWrite(yellowLED, LOW);
            digitalWrite(greenLED, HIGH);
            break;
       }
       client.println("HTTP/1.1 200 OK");
       client.println("Content-Type: text/html");
       client.println("Connection: close");
       client.println();
       client.println("<!DOCTYPE html>");
       client.println("<html>");
       client.println("<head>");
       client.println("<title>Arduino Controller</title>");
       client.println("</head>");
       client.println("<body>");
       client.println("<h2>Arduino Controller</h2>");

       client.print("<a href=\"");
       client.println("/?1\">Activate Red LED</a><br />");

       client.print("<a href=\"");
       client.println("/?2\">Activate Yellow LED</a><br />");

       client.print("<a href=\"");
       client.println("/?3\">Activate Green LED</a><br />");
       client.println("</body>");
       client.println("</html>");
       delay(10);
       client.stop();
     }
   }
 }
}
```

2. Save the sketch as **sketch2202**.
3. Click the Upload icon to verify, compile, and upload the sketch to your Arduino.
4. Make sure that the Arduino is plugged into the network, and then open the serial monitor tool in the Arduino IDE.
5. Press the Reset button on the Ethernet Shield to reset the server.
6. Look at the serial monitor output to see the IP address assigned to the Arduino.
7. Open a browser in a workstation on the network and connect to the URL:

   ```
   http://<ipaddress>/?0
   ```

   where *<ipaddress>* is the numeric IP address shown in the serial monitor output. The 0 will not activate any of the LEDs, but will return the web page that the sketch generates. This should display the web page shown in Figure 22.4.

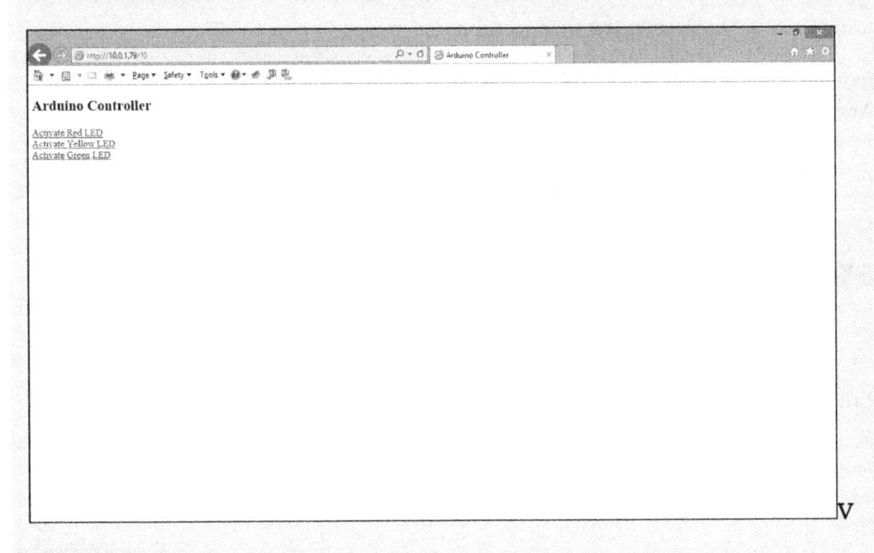

FIGURE 22.4
The main Arduino Controller web page.

8. Click one of the links to activate an LED.

When you click a link on the web page, the associated LED should light up on the Arduino. Clicking each link in the web page sends a new request to the web server. For example, clicking the Activate Red LED link sends the following request:

```
http://10.0.1.79/?1
```

The sketch reads the 1 value, which triggers the switch statement to run the `digitalWrite` functions to activate the red LED and deactivate the yellow and green LEDs.

Instead of using LEDs, you can connect anything to the digital interface pins, such as motors and relays. That enables you to control just about anything from a web client!

Summary

The Ethernet Shield for the Arduino allows you to use your Arduino as a web server on the network. HTTP provides a simple protocol that you can easily work with in your sketches to both send sensor data to remote clients, as well as read requests from remote clients to change the state of interfaces on your Arduino. Remote web clients send HTTP request messages to the Arduino, and the Arduino sends HTTP response messages. To retrieve sensor data, just embed the sensor data in the HTTP response, using the HTML language to format the web page output. To control Arduino interfaces, the client must embed a command inside the HTTP request, and the Arduino sketch much be able to decode the command in the request.

The next hour covers another important feature in Arduino sketches: storing data. The EEPROM included in the Arduino makes for a handy way to store data, but it's limited in size. To store larger amounts of data, you can use a shield that incorporates an SD card interface. The next hour walks through just how to read and write data using the SD card interface.

Workshop

Quiz

1. What HTTP response code indicates that the request was successfully processed?

   A. 300

   B. 200

   C. 500

2. You must send an HTTP header before you can send HTML data. True or false?

3. Which `EthernetClient` method should you use to check whether the remote client has connected to the server?

Answers

1. B. The HTTP server must return a 200 response code to indicate the HTTP request was received and processed correctly.
2. True. The HTTP response requires that you send an HTTP response header before you can send the HTML data to create the web page for the client.
3. After assigning the `server.available` method output to an `EthernetClient` object, you can use the `connected` method to determine whether the client is connected to the server, and the `available` method to determine whether there is data sent by the client.

Q&A

Q. How many digital inputs can you control on the Arduino Uno from a web server sketch?

A. Because the Ethernet Shield requires digital interface pins 10, 11, 12, and 13 to communicate to the Arduino, you can only use pins 0 through 9 for your sketches.

Q. Can multiple remote clients connect to the Arduino web server at the same time?

A. Yes. The `server.available` method will continue to listen for incoming connections and will accept connections from multiple remote clients at the same time. This requires you to be careful when controlling the Arduino from a remote client. Remember that more than one client can connect and send commands at the same time.

HOUR 23
Handling Files

What You'll Learn in This Hour:

- How to use SD cards in an Arduino
- How to use the Arduino SD library
- How to read and write files using an SD card

If you need to store data for long term on your Arduino, you have a couple of options. The EEPROM memory can store data, but there's a limit to how much space is available. Another option is to use one of the shields that contains an SD card reader, and use the SD library in Arduino to write files to it. This hour shows you just how to do that.

What Is an SD Card Reader?

With the boom in digital camera use, the ability to store large amounts of data in a small footprint became a necessity. One solution was the Secure Digital (SD) memory card format. It was developed in 1999 by a consortium of digital multimedia companies (SanDisk, Matsushita, and Toshiba) as a standard for storing data on a nonvolatile memory card for portable devices, such as digital cameras, mobile phones, and tablet computers. Over the years, there has been some tweaking done to the standard, including the physical size of the cards. At the time of this writing, there are currently three different physical sizes of SD cards:

- Standard SD (32mm × 24mm)
- MiniSD (21.5mm × 20mm)
- MicroSD (11mm × 15mm)

Most of the SD card interfaces available for the Arduino utilize the MicroSD card size. However, most MicroSD card packs include a converter that allows you to plug a MicroSD card into a standard SD card reader, often found on workstations and laptops.

Besides the physical changes, as the standard matured there have also been changes in the storage capacity of SD cards. As you can expect, the storage capacity has greatly increased from the original standard of 1MB. Table 23.1 shows the different capacity classes of SD cards.

TABLE 23.1 SD Card Capacity Classes

Classification	Storage Size	Format
SD Standard Capacity (SDSC)	1MB to 2GB	FAT16
SD High Capacity (SDHC)	2GB to 32GB	FAT32
SD Extended Capacity (SDXC)	32GB to 2TB	exFAT

All SD cards are preformatted by the manufacturer using one of three Windows disk formats. The exFAT format is a proprietary Microsoft format that has not been released to the public at the time of this writing. Because of that, the current Arduino SD library cannot interface with SDXC cards. That means the maximum size of SD card you can use with your Arduino is 32GB.

WATCH OUT!

Reformatting SD Cards

You may be tempted to purchase an SDXC card and try to reformat it to the FAT32 file system. Unfortunately, that technique has had mixed results with the current Arduino SD readers and library functions. Currently, it's not recommended to reformat SDXC cards. Instead, use an SDSC or SDHC card less than 32GB in size.

Also, if you must reformat an SDHC card, it's recommended to use the FAT16 format for cards up to 4GB in size.

The FAT16 file system format requires that you name files using the old 8.3 naming standard: an eight-character (or less) name, followed by a period, followed by a three-character (or less) file extension. Because of that requirement, the SD Library handles all files using the 8.3 format, even if you're using an SD card formatted with the FAT32 file system format.

When purchasing an SD card, besides the physical size and the storage capacity, there's one other feature that you need to take into consideration. There are also different classes of write speeds for the SD cards:

- **Class 2:** 2 MBps
- **Class 4:** 4 MBps
- **Class 6:** 6 MBps
- **Class 10:** 10 MBps

The Arduino SD card interfaces can use SD cards from any of these speed classes.

WATCH OUT!

SD Card Durability

The downside to SD cards is that they have a limited life span. The electronics contained in the card that store the data have a limited number of times the card can be written to, as well as how long the data can be retained on a card. Most SD cards are manufactured to retain data for 10 years, and can be written to about 100,000 times. After that, the card may still work, but may become unreliable.

SD Cards and the Arduino

The great thing about using an SD card in your Arduino project is that it provides an easy way to share data between your Arduino and a workstation. Because SD cards use the standard Microsoft FAT16 or FAT32 file system format, you can read and write to them from both Windows and OS X workstations (assuming your workstation has an SD card reader).

Currently, two official Arduino shields support SD cards on the Arduino:

- Ethernet Shield
- WiFi Shield

Figure 23.1 shows the Arduino Ethernet Shield with a MicroSD card in it.

FIGURE 23.1
The Arduino Ethernet Shield with the SD card reader.

Along with those two shields, the Arduino Ethernet device also includes an SD interface. All three platforms accept the MicroSD card size.

Besides the standard Arduino shields, some third-party companies have created SD card interfaces, such as the Adafruit Data Logging Shield, the Wave Shield, and the Micro-SD breakout board. For projects that just require a simple SD card interface, the Micro-SD breakout board contains everything you need in one small package.

All Arduino SD cards use the standard SPI pins to communicate with the SD card interface. For the Arduino Uno, that's digital pins 11, 12, and 13. For the Mega, it's pins 50, 51, and 52.

Besides those pins, the SD card readers also need to use an interface for the Select pin. By default, the Uno uses digital interface 10 as the Select pin (SS) to tell the SD card interface when the Arduino is communicating with it. However, with the Ethernet and WiFi shields, that's been moved to digital interface 4.

The SD Library

The Arduino IDE contains the SD library for interfacing with SD cards. You can use this library whether you're using an SD reader contained in a shield, a breakout card, or the Arduino itself.

There SD library consists of two classes of methods:

- **`SD` class:** Provides access to the SD card and manipulate files and directories on the card.
- **`File` class:** Provides access for reading and writing individual files on the SD card.

The following sections detail the methods contained in each of these classes.

The `SD` Class

The `SD` class provides methods to initialize the SD card interface and for working with files and directories from a high-level, such as opening a file or creating a directory. Table 23.2 shows the methods that are available in the `SD` class.

TABLE 23.2 The `SD` Class Methods

Method	Description
`begin(`*`pin`*`)`	Initializes the SD card (can specify a separate SS pin if not using the standard).
`exists(`*`file`*`)`	Tests if a file or directory exists.
`mkdir(`*`path`*`)`	Creates a directory; can also create intermediate directories (`mkdir /a/b/c`).

Method	Description
`open (filepath, mode)`	Opens a specified file using `FILE_READ pr FILE_WRITE`. If the file doesn't exist, `FILE_WRITE` will create it. Returns `File` object.
`remove(filepath)`	Removes a file.
`rmdir(path)`	Removes a directory (must be empty).

You'll need to use the `begin` method before you can do anything with the SD card. The SD library needs to know the Arduino interface pin used to control access to the SD card reader (called the *SS pin*). For most SD card interfaces, that's pin 10, but for the Ethernet and WiFi shields, pin 10 is used for the network connection, so the SS pin is moved to pin 4. Just specify the pin used as the parameter to the `begin` method:

```
SD.begin(4);
```

After initializing the SD interface, you use the `open` method to open a specific file for reading or writing. The `open` method returns a file handle to refer to a specific file. Since Arduino version 1.0, you can have multiple file handles open at the same time.

```
SD.begin(4);
File myFile;
myFile = SD.open("test.txt", FILE_WRITE);
```

The `open` method must specify both the name of the file to access, and the mode to open it (either `FILE_READ` or `FILE_WRITE`). After you have a file handle open, you can interact with the file using the methods in the `File` class.

The `File` Class

The `File` class provides the methods to read data from files, write data to files and a few methods for getting file information and handling file pointers. Table 23.3 shows these methods.

TABLE 23.3 The `File` Class Methods

Method	Description
`available`	Checks whether bytes available for reading.
`close`	Closes file; saves any unsaved data.
`flush`	Ensures that all bytes written to the file are physically saved.
`name`	Returns the name of file or folder.
`peek`	Reads a byte without advancing to the next one.
`position`	Gets the current position in the file (location of next read/write).

Method	Description
`print(data, BASE)`	Prints data to file. Numbers are printed as digits using the optional *BASE* specified.
`println(data, BASE)`	Prints data to file, followed by carriage return and newline.
`seek(pos)`	Seeks a new position in the file.
`size`	Returns the size of the file.
`read`	Reads a byte from the file. –1 if none available.
`write(data)`	Writes byte, char, or string to file.
`write(buf, len)`	Writes an array of characters or bytes.
`isDirectory`	Returns true if file is a directory.
`openNextFile`	Returns name of next file or folder in a directory.
`rewindDirectory`	Sets pointer back to first file or folder in a directory.

The `File` class uses the file handle that the `SD begin` method creates when opening a file. That process looks like this:

```
SD.begin(4);
File myfile = SD.open("test.txt", FILE_WRITE);
myFile.println("This line is saved in the file");
myFile.close();
```

The `myfile` variable contains the file handle that points to the `myfile.txt` file on the SD card. You can then use the `print`, `println`, or `write` methods to write data to the file, or the `read` method to read data from the file.

WATCH OUT!

Flushing Data

When you write data to the SD card file, the Arduino library stores the data in an output buffer. The data isn't actually written to the file until you use either the `flush` or `close` methods. To prevent corruption, be careful that you don't remove the SD card until your sketch runs the `close` method to close the file.

Interfacing with the SD Card

Now that you've seen the classes and methods that you need to work with files, it's time to start writing some sketches to use them. The following sections walk through the steps you need to create, read, update, and delete files and folders on your SD card. To run these examples, make sure that you have a formatted SD card in your SD card reader.

Writing to Files

Thanks to the SD library, writing data to a file on an SD card from your Arduino sketch is a simple process:

```
#include <SD.h>
File myFile;
int value;
unsigned long time;

void setup() {
   pinMode(10, OUTPUT);
   SD.begin(4);
}

void loop() {
   myFile = SD.open("test.csv", FILE_WRITE);
   time = millis();
   if (myFile) {
      value = analogRead(0);
      myFile.print(time);
      myFile.print(",");
      myFile.println(value);
   }
   myFile.close();
   delay(60000);
}
```

At the current time, a quirk in the SD library requires that you set digital interface 10 for output mode, even if your SD card reader doesn't use pin 10 for the SS interface. In this example, the sketch opens a file named test.txt on the SD card. If the file doesn't exist, the `open` method creates it.

The `loop` function first opens the file, writes the sensor data to it, and then closes the file. This helps ensure that the file is properly closed if you turn off the Arduino between loop iterations.

Reading Files

Reading files from the SD card can be a little tricky. You must know what data to expect when you read the data using your sketch. Fortunately, a few tricks can help with that process.

One trick is to pick a standard format to save the data in the file. The most popular way to save multiple data values in a text file is to use the comma-separated values (CSV) format. This format separates multiple values using a comma, as follows:

```
1002001,30,23
```

Each line in the CSV file represents a separate data record, often related to a separate reading iteration. The separate values within the data record are separated using commas. When you read the value from your sketch, you can use standard string functions (see Hour 8, "Working with Strings") to separate out the different values.

The other tricky part to this process is that the `File read` method can only read 1 byte of data at a time. You'll need to use more of the string functions to concatenate the data bytes to create the full string and to parse the strings into meaningful data:

```
#include <SD.h>
File myFile;
char ch;

void setup() {
   pinMode(10, OUTPUT);
   Serial.begin(9600);
   SD.begin(4);

   myFile = SD.open("test.csv", FILE_READ);
   if (!myFile) {
     Serial.println("Unable to open file");
   } else
   {
     while(myFile.available()) {
        ch = myFile.read();
        Serial.write(ch);
     }
     Serial.println("No more data available");
     myFile.close();
   }
}

void loop() {
}
```

The sketch tries to open the test.csv file, and if successful, it reads the file 1 byte at a time, displaying each character in the serial monitor until there is no more data in the test.csv file to read.

Working with Folders

By default, all the files that you create on the SD card are placed in the root folder of the drive. If you're using the same SD card for multiple projects, of if you need to store multiple types of data on the same SD card, you'll want to organize the data into folders.

The SD class provides methods for creating, listing, and deleting folders. First, let's go through an example of creating a couple of folders:

```
#include <SD.h>
void setup() {
   pinMode(10, OUTPUT);
   SD.begin(4);
   SD.mkdir("/test1");
   SD.mkdir("/test2");
   SD.mkdir("/test3/test4");
}

void loop() {
}
```

When the path specified in the `mkdir` method contains subfolders, the SD library will create any parent folders. If you remove the SD card and place it in a workstation or laptop SD card reader, you'll see the test3 folder at the root level, with the test4 folder under the test3 folder.

Now that you have some folders, here's an example that lists the folders:

```
#include <SD.h>
File rootFolder;
File myFile;
void setup() {
   Serial.begin(9600);
   pinMode(10, OUTPUT);
   SD.begin(4);
   rootFolder = SD.open("/", FILE_READ);
   while(true) {
      myFile = rootFolder.openNextFile();
      if (!myFile) {
         break;
      }
      if (myFile.isDirectory()) {
         Serial.println(myFile.name());
      }
   }
}

void loop() {
}
```

After you have the folders created, you can store your data files in them. To do that, just use the folder in the filename in the open method:

```
SD.open("/folder1/test.txt", FILE_READ);
```

Notice that you use forward slashes to separate the folder name from the filename, even though the SD card is formatted using a Windows file system.

Finally, if you want to remove folders, you use the rmdir method:

```
#include <SD.h>
void setup() {
   pinMode(10, OUTPUT);
   SD.begin(4);
   SD.rmdir("/test1");
}
void loop() {
}
```

After running this example, place the SD card into a workstation or laptop SD card reader to view the folders. You should see that the test1 folder is gone.

Storing Sensor Data

Let's go through an example that stores data from a temperature sensor in a file on the SD card. You can then remove the SD card from the Arduino when you're done and read it on a standard workstation or laptop that contains an SD card reader.

▼ TRY IT YOURSELF

Temperature Logger

In this exercise, you'll create a sketch that polls a TMP36 temperature sensor every minute and writes the values to an SD card. First you need to build a quick circuit to interface the TMP36 analog temperature sensor to the analog interface 0 pin on the Arduino:

1. Plug the TMP36 into a breadboard so that each lead is on a separate rail, and the flat side of the TMP36 is on the left side.
2. Connect the top lead of the TMP36 to the 5V pin on the Arduino.
3. Connect the bottom lead of the TMP36 to the GND pin on the Arduino.
4. Connect the middle lead of the TMP36 to the analog 0 interface on the Arduino.

That's all the hardware you need for this exercise. Next, you need to create the sketch code to read the sensor, convert it to a temperature, and then store the value in a data file on the SD card every minute. Just follow these steps:

1. Open the Arduino IDE, and enter this code into the editor window:

```
#include <SD.h>
File myFile;
unsigned long int time;
int tempF;

void setup() {
   Serial.begin(9600);
   pinMode(10, OUTPUT);
   SD.begin(4);
}

void loop() {
   myFile = SD.open("temp.csv", FILE_WRITE);
   time = millis();
   tempF = getTemp();
   myFile.print(time);
   myFile.print(",");
   myFile.println(tempF);
   myFile.close();
   Serial.print(tempF);
   Serial.print(" recorded at time ");
   Serial.println(time);
   delay(60000);
}

int getTemp() {
   int output;
   float voltage, tempC, tempF;
   output = analogRead(A0);
   voltage = output * (5000.0 / 1024.0);
   tempC = (voltage - 500) / 10;
   tempF = (tempC * 9.0 / 5.0) + 32.0;
   return int(tempF);
}
```

2. Save the sketch as **sketch2301**.
3. Click the Upload icon to verify, compile, and upload the sketch to your Arduino unit.
4. Open the serial monitor, then let the sketch run for 5 or 10 minutes. Figure 23.2 shows the output you should see in the serial monitor.
5. Remove the Arduino from the power source (either the USB cable or external power).

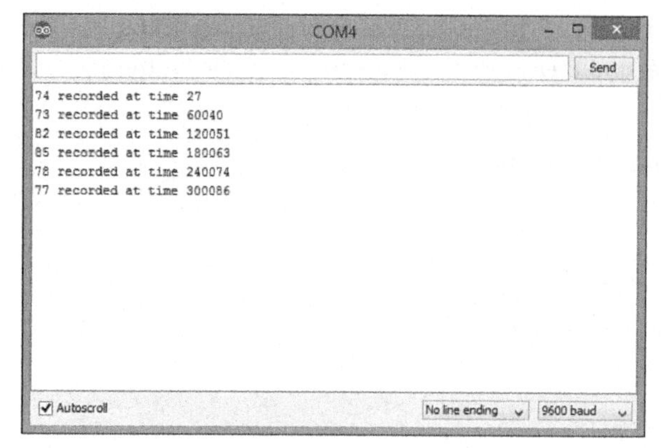

FIGURE 23.2
Output from the temperature logger sketch.

6. Eject the SD card from the SD card reader on the Arduino (or shield device), and insert it into the SD card reader on a workstation or laptop.
7. View the contents of the SD card using the file browser on your OS (such as Finder on OS X, or File Explorer on Windows).
8. Open the temp.csv file using either a text editor, or if available, a spreadsheet application such as Excel or LibreOffice Calc. Figure 23.3 shows what the data looks like in Excel.

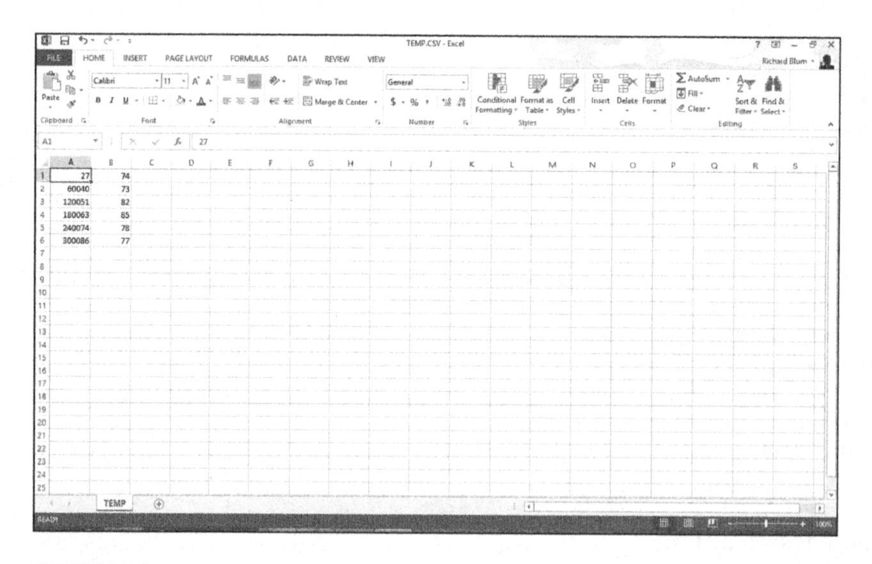

FIGURE 23.3
Viewing the temp.csv log file using Excel.

When you open the data file, you should see the sensor value.

Summary

This hour discussed how to use the SD card readers available on some shield devices as well as the Arduino Ethernet device. You use the SD library to interface with the SD card. The SD library allows you to both read and write to files on the card, as well as create and remove folders.

The next hour wraps up this book by showing how to create a complete Arduino project from start to finish. A few different steps are required to get a complete project up and running properly. The last hour walks through all of these steps to give you an idea of how to create your own Arduino projects.

Workshop

Quiz

1. What physical size of SD cards do most SD card readers available for the Arduino use?

    A. Standard SD

    B. MiniSD

    C. MicroSD

    D. SDXC

2. Data written by the Arduino on an SD card can be read from any workstation or laptop that supports an SD card reader. True or false?

3. What `File` method should you always use in your sketches to ensure that the data files aren't corrupted if the power is lost to the Arduino?

Answers

1. The Arduino Ethernet device, as well as the Ethernet Shield and WiFi Shield, all use the MicroSD cards for storing data.

2. True. The Arduino SD library can write data using the standard FAT16 or FAT32 file formats, allowing them to be written on any Windows, OS X, or Linux system.

3. You should always use the `close` method to properly close the data files when not in use. If the power is lost while the file is closed, you won't lose any data.

Q&A

Q. **Can I have more than one file open on an SD card at the same time?**

A. Yes. The current version of the SD library allows multiple files to be open at the same time.

Q. **Can I use a real date and time value when logging sensor data into a data file?**

A. Yes, but it's tricky. By default, the Arduino doesn't include a real-time clock, but there are solutions to that. Adafruit provides a hardware real-time clock add-on that you can add to your Arduino circuit to track the current date and time. Also, there is a DateTime software library available for the Arduino that can keep track of the date and time after its been set when the sketch starts.

HOUR 24
Prototyping Projects

What You'll Learn in This Hour:

- How to plan your Arduino project
- How to create a schematic
- How to design your Arduino sketch
- How to build a prototype circuit

The focus of this book has been on the programming side of Arduino, but creating a complete Arduino project takes a little more than just programming skills. As you've seen throughout the previous hours, you usually need to also build an electronic circuit to go along with your sketch code. This final hour discusses how to plan your Arduino projects to coordinate all the hardware and software pieces that you need to build your Arduino project.

Determining Project Requirements

The first step in any Arduino project is to create a list of the project requirements. When you start a project, you probably have some idea of just what you want it to do, but you might not have thought yet about just how the Arduino will do it. That's where project planning comes in.

Project planning consists of trying to document as many details about the project that you can find out up front, before you dive into the coding and building hardware. To plan out the project, you want to first ask yourself a few questions to determine the project requirements. Common planning questions to ask include the following:

- What types of data does the project need to monitor?
- What type of equipment does the project need to control?
- Does the sketch need to perform any calculations on the data?

- Does the project need to display any information?
- Does the data need to be saved?
- Does the project need to connect to a network?

As you think through these questions, you should start getting an idea of just what your Arduino project will look like and what components you need to create it.

▼ TRY IT YOURSELF

Planning an Arduino Project

To demonstrate how to plan a project, let's walk through the steps of designing and building a temperature monitor. The temperature monitor will allow us to preset a temperature, and then indicate if the room temperature is too hot or too cold.

Following our project plan, we need to start asking some questions:

1. First, because we need to monitor a temperature, we need to find a temperature sensor to use. For this project, we'll use the TMP36 analog temperature sensor.
2. The project will need a way to set a preset temperature level. For this project, we'll use a potentiometer to match the analog output voltage level of the temperature sensor for the temperature to set.
3. The project will need a way to display information to let us know whether the room temperature is too cold or too hot. For that, we'll use five LEDs:
   - A green LED will indicate when the temperature is within 5 degrees (plus or minus) of the preset temperature.
   - Two separate yellow LEDs will show when the temperature is between 5 and 10 degrees off (one LED for too high, and another for too low).
   - Two red LEDs to show when the temperature is more than 10 degrees too high or too low.

   Figure 24.1 shows the layout of the LEDs for the project.
4. The project won't save any data, and it won't use the network to communicate with any remote devices.
5. We will use the serial monitor to make sure the monitor is working correctly while we develop the sketch, but it won't be necessary for when we use the monitor live in production.

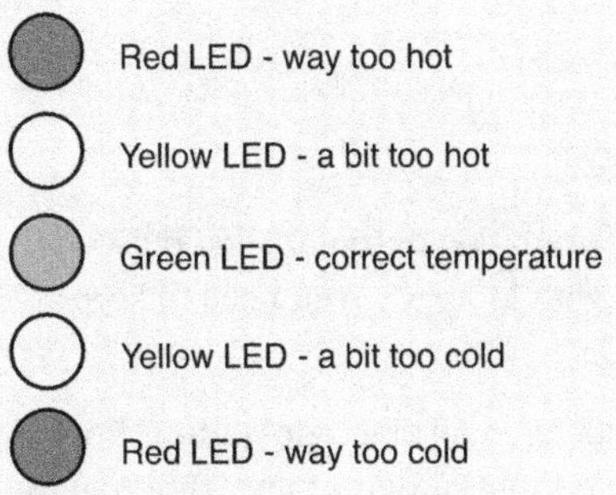

Red LED - way too hot

Yellow LED - a bit too hot

Green LED - correct temperature

Yellow LED - a bit too cold

Red LED - way too cold

FIGURE 24.1
The temperature monitor project LED layout.

Now that we've worked out the project planning, the next step is to determine just what the Arduino requirements will be.

Determining Interface Requirements

After you determine the project requirements, the next step is to map out just what Arduino interfaces are required. Remember, only a limited number of analog and digital interfaces are available on the different Arduino models, so the number of interfaces your project requires may also determine the Arduino model you need to use (or even if you need to use more than one Arduino unit for the project). This section discusses how to determine the interface requirements for both the analog and digital interfaces.

Analog Interfaces

You use analog interfaces for not only analog input. On the Arduino Uno, you also need to consider whether your project uses the I2C protocol to communicate with other devices. If your project needs to use I2C communication (such as to display information on an LCD shield), you'll have two fewer analog inputs available (A4 and A5).

For this exercise project, you just need to use two analog interfaces:

- One for the temperature sensor output
- One for the potentiometer output

After you identify how many analog interfaces you need, the next step is to assign them to interfaces. For this exercise, you use analog interface A0 for the temperature sensor output and analog interface A1 for the potentiometer output. That way, analog interfaces A4 and A5 will still be available if you decide to use an LCD shield later on.

Digital Interfaces

Digital interfaces can be used for digital input, digital output, or analog output (called pulse-width modulation (PWM), as described in Hour 19, "Working with Motors"). You need to keep track of which digital interfaces need to operate in which mode.

Besides those options, two digital interfaces (pin 0 and 1) can be used for the serial output from the Arduino. If you use the serial monitor in the Arduino IDE to input or output text, that means you can't use digital interface pins 0 or 1.

Also, the Serial Peripheral Interface (SPI) communication uses digital interface pins on the Arduino. If your sketch communicates with a sensor that uses SPI, you need to know which those ports are. For the Uno and Due models, you need to reserve digital interface pins 10, 11, 12, and 13 if you use SPI. For the Mega model, you need to reserve digital interface pins 50, 51, 52, and 53.

WATCH OUT!

The ICSP Header

Most Arduino models also support SPI communication using the separate in-circuit serial programming (ICSP) header. However, the SPI pins on the digital interface are still reserved and must be avoided if you're using the ICSP header pins.

Besides SPI, you also need to watch for I2C communication. While the Uno model uses analog pins for I2C, the Leonardo uses digital interface pins 2 and 3, and the Due and Mega models use digital interface pins 20 and 21.

For this project, you need to use five digital output interfaces, one for each LED you need to control:

- Interface 2 for the low-temp red LED
- Interface 3 for the low-temp yellow LED
- Interface 4 for the green LED
- Interface 5 for the high-temp yellow LED
- Interface 6 for the high-temp red LED

Because you use all the digital interfaces in the project for output, you don't need to track input or PWM interfaces. If you're creating a more complex project that uses those features, it may help to create a table that shows all the digital interface, which mode they use, and what they're connected to.

Listing Components

The next step in the project is to determine what electronic components are required. This can consist of the following:

- Sensors
- Output devices (LEDs, LCDs)
- Switches
- Motors
- Auxiliary components (resistors, capacitors, transistors)
- Power components

It's often the little things in a project that are overlooked and cause problems when it's time to build the hardware. Don't forget the small items, such as resistors to limit the current going through LEDs, when listing out the component requirements.

As part of the component list, I like to also include the power requirements for the project. Some components require 5 volts to operate, whereas others require 3.3 volts. If you're working with motors, you also need to use an external power source to provide more voltage to run the motor, such as a 9V battery.

For this project, you need the components shown in Table 24.1.

TABLE 24.1 Project Components

Component	Quantity
Red LED	2
Yellow LED	2
Green LED	1
1K-ohm resistor	5
TMP36 sensor	1
1K potentiometer	1

Both the TMP36 sensor and the 1K potentiometer can use either 3.3 volts or 5 volts to operate, which you can get directly from the Arduino, so you don't need to worry about any external power sources. After you determine your component list, you can start planning how to connect them. To do that, you'll want to create a schematic.

Creating a Schematic

You'll want to map out a schematic of the project hardware so that you can determine how to connect the components. Software packages are available that can do that for you, or you can just map out the schematic freehand. Figure 24.2 shows an example of drawing the schematic freehand.

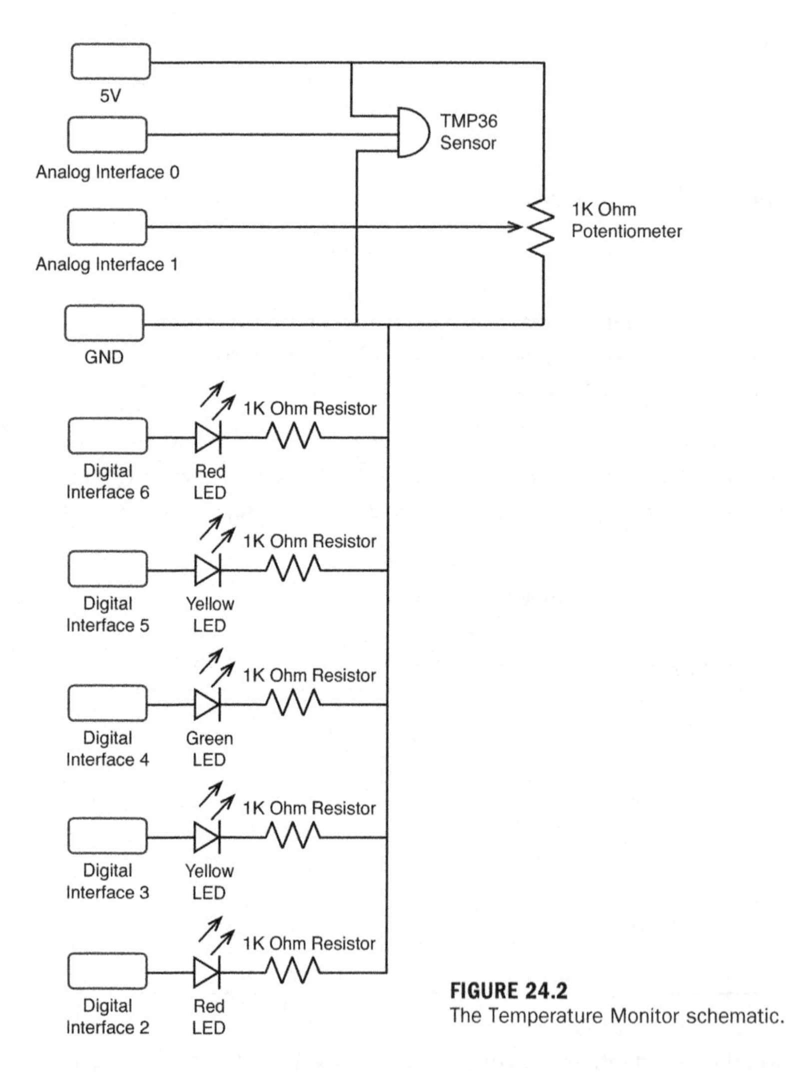

FIGURE 24.2
The Temperature Monitor schematic.

When you map out the circuit schematic, make sure to cover all the connections to the Arduino.

Creating the Breadboard Circuit

After you have the schematic drawn out, you can start plugging components into a breadboard to wire the circuit. Breadboards are great tools for temporarily creating your Arduino circuits. Because the connections are temporary, it's easy to change things around until you get them right.

TRY IT YOURSELF ▼

Wiring the Temperature Monitor

To create the temperature monitor circuit, follow these steps:

1. Place the five LEDs on a breadboard so that they straddle the middle divider. Place the long lead of the LED toward the left side of the breadboard. Line them up so that there's a red LED at the top, followed by a yellow LED, the green LED, the other yellow LED, and finally, the other red LED.
2. Connect a 1K-ohm resistor from the short lead of each LED to a common rail area on the breadboard.
3. Connect that common rail area to the GND pin on the Arduino.
4. Connect the five digital interface pins on the Arduino to the five LEDs using the sequence:
    - Interface 2 connects to the bottom red LED.
    - Interface 3 connects to the bottom yellow LED.
    - Interface 4 connects to the green LED.
    - Interface 5 connects to the top yellow LED.
    - Interface 6 connects to the top red LED.
5. Connect a wire from the 5V interface on the Arduino to another common rail area on the breadboard.
6. Place the TMP36 sensor on the breadboard so the flat side points to the left and so that each sensor lead is plugged into a separate rail area.
7. Connect the top pin of the TMP36 sensor to the 5V rail.
8. Connect the bottom pin of the TMP36 sensor to the GND rail.
9. Connect the middle pin of the TMP36 sensor to the Arduino A0 interface.
10. Place the potentiometer on the breadboard so that each lead plugs into a separate rail area.

11. Connect one outer lead of the potentiometer to the 5V rail on the breadboard.
12. Connect the other outer lead of the potentiometer to the GND rail on the breadboard.
13. Connect the middle lead of the potentiometer to the A1 interface on the Arduino.

That was a lot of wires to connect! There are nine separate wires going between the Arduino and the breadboard—the five digital interfaces, two analog interfaces, the GND interface, and the 5V interface. Figure 24.3 shows how things should look.

FIGURE 24.3
The finished project wiring.

With the hardware setup complete, you can move on to working on the sketch design.

Designing the Sketch

With the hardware complete, it's time to turn your attention to the software that will control the project. Just like planning out the project, you'll want to plan out how you want the sketch to work before you start writing any code. That will help when it does come time to write code.

Things you need to think about while planning the sketch include the following:

- Which Arduino libraries are required
- What variables and constants you will define

- What code needs to be in the `setup` function
- What code needs to be in the `loop` function
- Whether any extra functions are required

It's important to remember that in the Arduino, code inside the `setup` function is only run at startup, and any code inside the `loop` function is run in a continuous loop. If you need to initialize interface pins, serial interfaces, or the network interface, that code goes in the `setup` function area. When you want to monitor sensors, that code most often goes in the `loop` function area so that it continually runs and updates the output with new sensor data.

For this project, you need code for the following things:

- Initialize the serial monitor
- Initialize the digital interface inputs
- Retrieve the current temperature sensor value
- Retrieve the current potentiometer value
- Compare the two values and light the appropriate LED

You'll notice that the first two items involve initializing things: the serial monitor and the digital interfaces. The code for those features go in the `setup` function area because they need to run only once at the start of the sketch.

The remaining items on the list involve retrieving sensor values and producing an output. You place the code for those features in the `loop` function area so that they continually run. You also need to use a one second delay to pause the program between readings. That way you won't overrun the serial monitor with output during the tests!

The analog temperature sensor outputs a voltage based on the temperature it detects. To convert the voltage value to a real temperature, you need to create a separate function to help keep those calculations out of the `loop` function area. That's not required, but it helps keep the sketch code from getting too cluttered.

You'll also create a separate function to determine which LED lights to indicate the temperature. Because that requires a lot of `if-then` statements to compare the value ranges, it will help to keep that code out of the `loop` function area as well.

Writing the Sketch

After you've mapped out the basic functions and features you need for your sketch, you can start coding the sketch.

▼ TRY IT YOURSELF

Creating the Temperature Sensor Code

To build the temperature sensor sketch, just follow these steps:

1. Open the Arduino IDE, and enter this code:

```
int temp;
int setting;

void setup() {
  pinMode(2, OUTPUT);
  pinMode(3, OUTPUT);
  pinMode(4, OUTPUT);
  pinMode(5, OUTPUT);
  pinMode(6, OUTPUT);
  Serial.begin(9600);
}

void loop() {
  int scale;
  scale = map(analogRead(A1), 0, 1023, 0, 254);
  temp = getTemp(analogRead(A0));
  setting = getTemp(scale);
  Serial.print("Temp: ");
  Serial.print(temp);
  Serial.print("  setting: ");
  Serial.println(setting);
  checkTemp(temp, setting);
  delay(1000);
}

int getTemp(int value) {
   float voltage, tempC, tempF;
   voltage = value * (5000.0 / 1024.0);
   tempC = (voltage - 500) / 10;
   tempF = (tempC * 9.0 / 5.0) + 32.0;
   return int(tempF);
}

void checkTemp(int temp, int setting)
{
  if (abs(temp - setting) < 5)
  {
    // temperature just right
    digitalWrite(6, LOW);
```

```
        digitalWrite(5, LOW);
        digitalWrite(4, HIGH);
        digitalWrite(3, LOW);
        digitalWrite(2, LOW);
      } else if (((temp - setting) > 0) && ((temp - setting) < 10))
      {
        // temperature a little too hot
        digitalWrite(6, LOW);
        digitalWrite(5, HIGH);
        digitalWrite(4, LOW);
        digitalWrite(3, LOW);
        digitalWrite(2, LOW);
      } else if (((temp - setting) > 0) && ((temp - setting) < 15))
      {
        // temperature way too hot!
        digitalWrite(6, HIGH);
        digitalWrite(5, LOW);
        digitalWrite(4, LOW);
        digitalWrite(3, LOW);
        digitalWrite(2, LOW);
      } else if (((temp - setting) < 0) && ((setting - temp) < 10))
      {
        // temperature a little too cold
        digitalWrite(6, LOW);
        digitalWrite(5, LOW);
        digitalWrite(4, LOW);
        digitalWrite(3, HIGH);
        digitalWrite(2, LOW);
      } else if (((temp - setting) < 0) && ((setting - temp) < 15))
      {
        // temperature way too cold!
        digitalWrite(6, LOW);
        digitalWrite(5, LOW);
        digitalWrite(4, LOW);
        digitalWrite(3, LOW);
        digitalWrite(2, HIGH);
      }
    }
```

2. Save the sketch as **sketch2401**.
3. Click the Upload icon to verify, compile, and upload the sketch to the Arduino unit.

The `setup` function defines the output digital interfaces and initializes the serial monitor port. The `loop` function retrieves the current temperature value and converts it to a temperature using the `getTemp` function. Likewise, it also retrieves the temperature setting output from the potentiometer and converts it to a temperature as well.

Because the potentiometer can output a wider range of values than the temperature sensor, I scaled the digital output of the potentiometer to a smaller range. That will make the potentiometer a little less sensitive as you rotate the wiper to match the temperature voltage output.

The `checkTemp` function compares the sensor output to the potentiometer output, and lights the appropriate LED. If the temperatures are within 5 degrees plus or minus, it lights the green LED.

Testing the Sketch

With both the hardware and software complete, you're ready to test the sketch. Because the sketch outputs the sensor and potentiometer values to the serial monitor, it's a snap to see what's going on.

▼ TRY IT YOURSELF

Viewing the Sketch Output

To run the sketch and view the output, follow these steps:

1. When you click the Upload icon in the Arduino IDE, the sketch will start running automatically. Open the serial monitor to view the output that's generated from the sketch.
2. Turn the potentiometer to change the voltage output. Try to match the setting output to the temperature output as shown in the serial monitor. Figure 24.4 shows the output you should see.

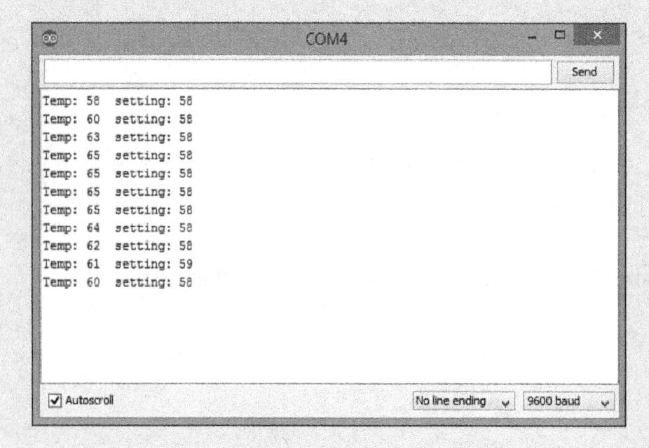

FIGURE 24.4
The serial monitor output from testing the temperature sensor.

3. When you get the temperatures to match, you should see the green LED light.
4. Leave the potentiometer setting alone, and then try to heat up and cool down the temperature sensor by holding it between your fingers and placing a bag of ice next to it.

As the sensor heats up, the top yellow and red LEDs should light in sequence. As the sensor cools down below the preset setting, the lower yellow and red LEDs should light. If you can't get the sensor to heat up or cool down enough to test the red LEDs, try changing the range values in the `checkTemp` function to smaller ranges.

Creating a Prototype Board

When you deploy your project in "real life," you most likely won't want to do it using a breadboard to hold the electronic components. The breadboard connections are temporary and may become loose over time or get accidentally pulled out.

The solution is to create a prototype circuit board that allows you to solder the components into place. There are two ways to create a prototype circuit:

- Use a circuit board design software package to create a circuit board
- Use a prototype circuit board

This section walks through these options for creating your prototype circuit.

Using a Prototype Board

A prototype circuit board provides some level of flexibility, with a more permanent wiring solution. It works similarly to a breadboard, by providing rails of interconnected sockets, but it works like a circuit board, with the interconnections being done on the circuit board by traces and the components being soldered into place.

You can use generic prototype circuit boards for your project, but what makes things even easier is that Adafruit created an Arduino-compatible prototype board.

The Proto Shield plugs into most Arduino models as a normal shield and provides a simple prototype circuit board on the shield itself. Figure 24.5 shows the Proto Shield.

With the Proto Shield, you can lay out the components for your project in the holes provided in the circuit board. Note that there are several rail sections where the holes are tied together on the circuit board. This allows you to easily share a common connection, such as all the ground connections or all the connections that go to the 5V pin on the Arduino.

FIGURE 24.5
The Adafruit Proto Shield.

The Proto Shield also provides all the Arduino interface pins directly on the board. You'll have to use small jumper wires to connect the LEDs to the digital interface pins and the sensor and potentiometer to the analog interface pins.

Creating a Circuit Board

A step above using the Proto Shield is to create your own custom circuit board. The advantage of creating your own custom circuit board is that you can place the components as you like on the board, which makes your project look more professional. Also, once you get your custom circuit board layout the way you want it, you can easily send that diagram to electronic companies to mass produce your project circuit board.

To create a circuit board from the schematic diagram requires some software. While there are some commercial circuit board design programs available, for hobbyists the most popular open source circuit board software is Eagle.

The Eagle package enables you to lay out your electronic components in a grid using a graphical tool. It then connects the component leads as required to satisfy the schematic that you provide.

After you generate the template for the circuit board, you can either print it out to etch into a printed circuit board, or you can send the file to a company that specializes in mass producing printed circuit boards. Many companies recognize Eagle-formatted circuit files.

Summary

This final hour walked through how to prototype your Arduino projects to get them ready for the real world. The first step is to plan just what you need for the project to accomplish and determine what Arduino features and shields you need. After planning the project, the next step is to create a parts list of the components you need for the project. With that in hand, you can start mapping out a schematic of how the project circuit needs to look. With your schematic complete, the next step is to build a prototype circuit on a breadboard. After you have the circuit completed, you're ready to start coding. You'll need to map out the code requirements, and then implement them in your Arduino sketch. Finally, with your prototype complete and working, you can move the components to a more permanent solution, either using a Proto Shield or by building your own custom circuit board.

Workshop

Quiz

1. What software enables you to create your own custom circuit boards that can be used to mass produce your Arduino project?

    A. The Proto Shield

    B. The Arduino IDE

    C. The Eagle software

    D. The Ethernet Shield

2. When mapping out digital and analog interface requirements, you must also think about the I2C and SPI connection requirements in your project. True or false?

3. When designing your sketch code, how do you determine which features go in the `setup` function and which ones go in the `loop` function?

Answers

1. The Eagle software enables you to create custom circuit board templates that you can then use to build your own circuit board or send off to have the circuit board mass produced.
2. True. The different Arduino models use different interfaces to implement the I2C and SPI communication ports. You'll need to be careful to avoid those interfaces if your project uses those protocols.
3. Features that only need to run once to initialize values or interfaces should go in the `setup` function. Features that need to run continuously should go in the `loop` function.

Q&A

Q. Shouldn't you design the schematic circuit first before you create a components list for the project?

A. I like to determine at least what major components are required for a project before trying to map out the schematic. That helps me organize what the schematic needs to look like. Most likely you'll find you need more components as you create the schematic, which you can then add to your components list.

Q. Is designing an Arduino project an iterative process; do you need to repeat any of the individual steps to get to the final design?

A. Yes, you can use an iterative process to designing an Arduino project, but that can get tricky. Sometimes for larger projects, you can start out by designing a stripped-down version of the project features, and then go back and add additional features. However, that can make designing the schematic more difficult as you need to incorporate more components as you go along.

Index

Symbols

Numbers

A

B

C

D

E

F

G

H

I

K–L

How can we make this index more useful? Email us at indexes@samspublishing.com

M

Q

R

T

U

V

W

Y

Z